MACHINE SHOP TRAINING

S.F. Krar   J.W. Oswald

SI Metric   Fourth Edition

McGRAW-HILL RYERSON LIMITED
Toronto   Montréal   New York   Auckland   Bogotá   Cairo   Guatemala
Hamburg   Lisbon   London   Madrid   Mexico   New Delhi
Panama   Paris   San Juan   São Paulo   Singapore   Sydney   Tokyo

Machine Shop Training, Fourth Edition

Copyright © McGraw-Hill Ryerson Limited, 1986
All rights reserved. No part of this publication may be reproduced, stored in a retrieval system, or transmitted, in any form, or by any means, mechanical, electronic, photocopying, recording or otherwise, without the prior written permission of McGraw-Hill Ryerson Limited.

ISBN 0-07-548951-1

    5 6 7 8 9 0 THB 5 4 3 2 1

Printed and bound in Canada

Canadian Cataloguing in Publication Data

Krar, S. F., date
   Machine shop training

Includes index.
ISBN 0-07-548951-1

1. Machine-shop practice. I. Oswald, J. W. (James William). II. Title.

TJ1160.K73 1986   670.42′3   C85-099820-4

MACHINE SHOP TRAINING

S.F. Krar   J.W. Oswald

SI Metric   Fourth Edition

PREFACE

This fourth edition of *Machine Shop Training* has been updated to include the suggestions received from time to time from teachers and other interested individuals.

Although Canada and the United States are committed to the use of the metric (SI) system of measurement because of its inherent simplicity, the move toward this change has been slower than anticipated. It will be many years before all machine tools will be converted to metric measurement or declared obsolete. Consequently the need for dual measurement systems (metric and imperial) still exists. This need is particularly important in the classrooms, where many of the students from elementary schools have been educated in the metric system of measurement only. These people will have to be retrained to go into the "inch" world, since most manufacturing companies (their future employers) have not followed the proposed metric trends as quickly as anticipated. Because of this problem, both metric and inch measuring systems have been included in this textbook. Teachers and students must decide whether or not they will learn only metric tools and measurement or both the metric and inch systems, based on the latest industrial and educational practices.

All sections of the former edition were carefully examined and updated, expanded, or shortened as was seen fit. To further prepare a student for a job after leaving school, two new sections have been added to this textbook. The chapter on **Careers** will acquaint the student with the opportunities in the machine tool field of employment and the various career opportunities related to this area of employment.

Job Planning has been included throughout the book to acquaint the student with the logical sequence required to produce a part. Over the years it has been found that a student will perform an operation because it was laid down in the instruction sheets or by the teacher. It is hoped that this section will make the student more self-reliant.

In order to simplify the task of both the student and the teacher, the following features are incorporated in this book:

1. The material is organized so that the teacher may select those topics most suitable for the project or to suit the individual differences of the students.

2. Each operation is explained in a step-by-step, easy to follow procedure.

3. Operations have been kept as brief as possible. Related information which is not actually part of the operation has been highlighted.

4. Many new photographs have been included to further clarify the text.

5. Colour and shading have been used extensively to emphasize important points and to make the illustrations more meaningful.

6. Questions are provided at the end of most chapters. These may be used either as homework assignments to prepare students for new operations or for review purposes.

In order to be successful in the machine shop trade, a person should be neat, develop sound work habits, and have a good knowledge of practical mathematics and print reading as it relates to machine shop work. Due to the ever-changing technology, a person should keep abreast of new developments by reading specialized texts, trade literature, and articles related to this exciting line of work.

CONTENTS

PREFACE

ACKNOWLEDGEMENTS

CHAPTER 1 THE EVOLUTION OF MACHINE TOOLS — 1

CHAPTER 2 COMMON MACHINE TOOLS — 4

CHAPTER 3 CAREERS — 8

CHAPTER 4 SAFETY — 14

CHAPTER 5 MEASURING TOOLS AND SYSTEMS — 18

CHAPTER 6 LAYOUT TOOLS — 35

CHAPTER 7 METALLURGY — 48

CHAPTER 8 HAND TOOLS — 64

CHAPTER 9 POWER SAWS — 85

CHAPTER 10 DRILL PRESSES — 96

CHAPTER 11 THE ENGINE LATHE — 118

CHAPTER 12 THE SHAPER — 186

CHAPTER 13 MILLING MACHINES — 199

CHAPTER 14 GRINDERS — 231

CHAPTER 15 COMPUTER AGE MACHINING — 247

CHAPTER 16 HEAT TREATING — 255

APPENDIX TABLES — 262

GLOSSARY — 267

INDEX — 271

ACKNOWLEDGEMENTS

The authors wish to express their sincere appreciation to Alice H. Krar for the countless hours she devoted to typing, proofreading, and checking the manuscript for this fourth edition. Her assistance has contributed greatly to the book's clarity and completeness.

We owe a special debt of gratitude to the many teachers, students, and industrial personnel who were kind enough to offer constructive criticism and suggestions for improving this popular text. As many suggestions as possible were incorporated in this edition in order to make this a more useful reference for both the student and the instructor.

We are also grateful to the following firms, who were kind enough to review sections of the manuscript and supply technical information and illustrations for this text:

American Iron and Steel Institute
American Superior Electric Co. Ltd.
Atlas Press Company
Bausch & Lomb
Bendix Corporation
Bridgeport Machines, Division of Textron Inc.
Brown & Sharpe Manufacturing Company
Butterfield Division, Union Twist Drill Co.
Canadian Acme Screw and Gear Ltd.
Carborundum Company
Cincinnati Lathe & Tool Co.
Cincinnati Milacron Inc.
Cincinnati Shaper Co.
Cincinnati Tool Co.
Clausing Corp.
Cleveland Twist Drill (Canada) Ltd.
Colchester Lathe Co.
Cushman Industries Inc.
Delmar Publishers

Delta File Works
DoAll Company
Eclipse-Pioneer
Elliott Machine Tools
Everett Industries Inc.
Excello Corp.
Firth-Brown Tools (Canada) Ltd.
Frontier Equipment
Charles A. Hines Inc.
Industrial Accident Prevention Association of Ontario
Inland Steel Corporation
Jacobs Manufacturing Co.
Kaiser Steel Corp.
Kostel Enterprises Ltd.
R.K. LeBlond Machine Tool Co.
Linde Division, Union Carbide Corporation
Lufkin Rule Co. of Canada, Ltd.
Monarch Machine Tool Co.
Morse Twist Drill and Machine Co.

Nicholson File Co. of Canada, Ltd.
Norton Co. of Canada, Ltd.
H. Paulin & Co. Ltd.
Pratt & Whitney Co. Inc.
Rockwell International
Shell Oil Co. of Canada Ltd.
South Bend Lathe, Inc.
Standard-Modern Technologies
Standco Canada Ltd.
Stanley Tools Division, Stanley Works
L.S. Starrett Co.

Steel Company of Canada, Ltd.
Sunbeam Equipment Corporation
Taft-Peirce Manufacturing Co.
United States Steel Corporation
Walker-Turner Division, Rockwell Manufacturing Co.
Weldon Tool Co.
Wells Manufacturing Corp.
A.R. Williams Machinery Company
J.H. Williams & Co.

CHAPTER 1
THE EVOLUTION OF MACHINE TOOLS

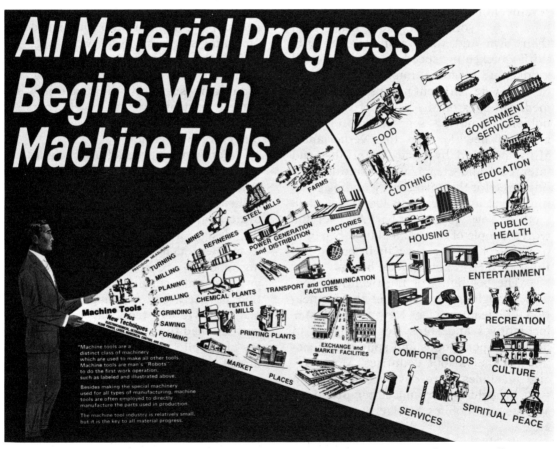

Courtesy DoAll Company

Four thousand years ago, the chieftains of northern Europe were just beginning to discard their stone spears and axes in favour of bronze weapons brought up the Danube and across the Bay of Biscay by people from the Mediterranean. Five hundred years later, in the mountains near the Caspian Sea, iron was being mined and wrought into ornaments for kings. Iron did not replace bronze in tools and weapons until about three thousand years ago, when the smiths learned how to harden and temper it. Once this happened, the age of iron and steel began. The iron-rimmed wheels of Assyrian chariots were followed by keen-bladed swords hammered out by the craftspeople of Damascus and by the armour, chain-mail, and steel-tipped arrows and spears that can be seen in a picture of the fourteenth-century Battle of Crécy.

About three hundred years ago, the Iron Age became the Machine Age. In the seventeenth century, people began to learn how to use sources of power other than their own and their animals' strength. With quickening speed, machines of all kinds were invented, improved, and used far and wide. Efficient pumps replaced the medieval ones on the banks of the Dutch polders, removing sea water to create acres of new land in the Netherlands. Mills, powered by wind, waterfalls, and later coal and steam, were continuously improved for the purpose of grinding grain into flour and operating looms to make cloth or saws to cut timber. All over the world, people of mechanical bent began to use their heads to save their backs and those of their fellow human beings.

The beginning of the twentieth century brought refinements in the application of steam power. Newer and faster locomotives and ships were designed to help transport goods and spur the economies of the various countries. More precise machines were developed, which in turn brought newer and better products, such as automobiles, refrigerators, planes, etc.

World War II brought on the need for greater production, and consequently more accurate and more efficient machines were developed. Production machines such as the turret lathe, automatic screw machines, and semi-automatic milling machines produced parts under the guidance of a machine operator. More specialized parts were produced by the machinist.

Around the middle of this century, numerical control of machines was introduced. Parts of greater accuracy were now produced more efficiently, and the need for better machines and technology grew. At the same time, new "space age" metals were produced and electro-machining processes were developed to machine them. Electronic devices replaced heavier and more costly electrical parts. Calculators came into being; nuclear power was harnessed and used in electrical power plants and ships. Computers were in their developmental stage and were very large. To satisfy the needs of the space program, computers were constantly refined and reduced in size until they have now become a part of our way of life, possibly having a greater effect on society than any other single invention.

Today most production machines are computer-controlled, and parts are assembled and welded by computer-controlled robots. Even design engineers and drafters are now assisted by computers in the design and drawing of the prints (computer-aided design, or CAD). "Working models" are now made on the computer to make certain that the parts function properly before being produced by machine. We can truly say that we have now reached the computer age.

In spite of this vast change in manufacturing, the basic processes such as drilling, milling, and turning are still used, and they always will be. In order to understand how a part is to be made, the designer and the machine or cell operator must understand the basic cutting processes (drilling, milling, and turning) and know the sequence that should be followed to produce any given part.

To operate today's machines, skilled technicians are indispensable. Such people are not just machine operators. They must be trained to approach and solve new problems as they arise. They must be capable of carrying out ideas and plans that call for the production of extremely intricate parts. In addition to possessing skill, they must have many other characteristics to be successful. Proper self-care, orderliness, good judgement, confidence, safe work habits, and attention to accuracy are some of the essentials required to become skilled at a craft.

CHAPTER 2
COMMON MACHINE TOOLS

Courtesy DoAll Company

Humanity's progress through the centuries has been governed by the types of tools it has developed. Cave dwellers used a bow-drill to make a hole probably as far back as prehistoric times (Fig. 2-1). By the early thirteenth century, people had learned to turn wood on a spring pole and treadle lathe (Fig. 2-2). Almost six centuries elapsed before two of the most important inventions in all of industrial history were developed. Eli Whitney introduced the principle of interchangeable-parts manufacture in the early nineteenth century. In the same decade, Charles Babbage conceived his calculating engine, a fine mechanism and an ancestor of the present-day computer.

As better tools were developed, the creative desire to further improve the basic tools and create new ones grew. In the first decade of the twentieth century, there was a rising tide of technical and

Fig. 2-1
The bow-drill was used by prehistoric people to drill a hole.

industrial advancement, especially in the electrical, automotive, hydraulic, and communications fields, and also in most other industries. Each new tool or development helped to produce more goods at a lower cost and further raise the standard of living of people throughout the world. Today, with the ever-increasing use of computers in the manufacturing process, we are on the threshold of developments which could have a greater effect on society than any other developments in the history of the human race.

The machine tool industry is divided into several different categories such as the general machine shop, the toolroom, and the production shop. The machine tools found in the metal trade fall into three broad categories:

1. *Chip producing machines*, which form metal to size and shape by cutting away the unwanted sections. These machine tools generally alter the shape of steel products produced by casting, forging, or rolling in a steel mill.
2. *Non-chip producing machines*, which form metal to size and shape by pressing, drawing, or shearing. These machine tools generally alter the shape of sheet steel or other metal products and granular or powdered materials.
3. *New generation machines*, which were developed to perform operations that would be very difficult, if not impossible, to perform on chip or non-chip producing machines. Electro-discharge and electro-chemical machines, for example, use either electrical or chemical energy to form metal to size and shape.

Fig. 2-2
The spring pole and treadle lathe was used in the thirteenth century to turn wood.

COMMON MACHINE TOOLS 5

Since this book is designed for the person beginning in the machine shop trade, the emphasis will be placed on basic operations and chip-producing machine tools.

A general machine shop contains a number of standard machine tools that are basic to the production of a variety of metal components. Operations such as turning, boring, threading, drilling, reaming, sawing, milling, and grinding are most commonly performed in a machine shop. Machines such as the drill press, engine lathe, power saw, milling machine, and grinder are usually considered the basic machine tools in a machine shop.

DRILL PRESS

The drill press, probably the first mechanical device developed by prehistoric people, is used primarily to produce round holes. Drill presses range from the simple hobby type to the more complex automatic and numerically controlled machines used for production purposes. The function of a drill press is to grip and revolve the cutting tool (generally a twist drill) so that a hole may be produced in a piece of metal or other material. Operations such as drilling, reaming, spot facing, countersinking, counterboring, and tapping are commonly performed on a drill press.

ENGINE LATHE

The engine lathe is used to produce round work. The workpiece, held by a work-holding device mounted on the lathe spindle, is revolved against a cutting tool which produces a cylindrical form. Straight turning, tapering, facing, drilling, boring, reaming, and thread cutting are some of the common operations performed on a lathe.

METAL SAW

Metal-cutting saws are used to cut metal to the proper length and shape. There are two main types of metal-cutting saws: the bandsaw (horizontal and vertical), and the reciprocating cut-off saw. On the vertical bandsaw the workpiece is held on the table and brought into contact with the continuous-cutting saw blade. It can be used to cut work to length and shape. The horizontal bandsaw and the reciprocating saw are used to cut work to length only. The material is held in a vise and the saw blade is brought into contact with the work.

SHAPER

The shaper is generally used for producing flat, curved, or angular surfaces on metal workpieces. The cutting tool moves back and forth in a horizontal plane across the face of the work held in a vise or fastened to the table. The work is moved across for successive cuts either by hand or automatic feed.

MILLING MACHINE

The horizontal milling machine and the vertical milling machine are two of the most useful and versatile machine tools. Both machines use one or more rotating milling cutters having single or multiple cutting edges. The workpiece, which may be held in a vise, fixture, or accessory, or fastened to the table, is fed into the revolving cutter. Equipped with proper accessories, milling machines are capable of performing a wide variety of operations such as drilling, reaming, boring, counterboring, spot facing, and producing flat and contour surfaces, grooves, gear teeth, and helical forms.

GRINDER

Grinders use an abrasive cutting tool to bring a workpiece to an accurate size and produce a high surface finish. In the grinding process, the surface of the work is brought into contact with the revolving grinding wheel. The most common types of grinders are the surface, cylindrical, cutter and tool, and the bench or pedestal grinder. *Surface grinders* are used to produce flat, angular, or contoured surfaces on a workpiece. *Cylindrical grinders* are used to produce internal and external diameters, which may be straight, tapered, or contoured. *Cutter and tool grinders* are generally used to sharpen milling cutters. *Bench and pedestal grinders* are used for offhand grinding and the sharpening of cutting tools such as chisels, punches, drills, and lathe and shaper tools.

SPECIAL MACHINE TOOLS

Special machine tools are designed to perform all the operations necessary to produce a single component. Some examples of special purpose machine tools are: gear-generating machines; centreless, cam, and thread grinders; turret lathes; automatic screw machines; and machining centres. The introduction of electro-discharge machining, electro-chemical machining, and electrolytic grinding have made it possible to machine materials and produce shapes that were difficult or impossible to produce by other methods. Numerical and computer control of machine tools has greatly increased production and improved the quality of the finished product.

COMPUTER-AIDED MANUFACTURING

The use of numerical control (NC) and computer numerical control (CNC) of machine tools in the mid-twentieth century has revolutionized machining processes by greatly increasing productivity and improving the quality of the finished product. Computer-aided design (CAD) has revolutionized the process that drafters and tool engineers use to design and test new products. The ability of computer-aided manufacturing (CAM) to accurately control the entire operation of machine tools has greatly improved productivity.

During the 1980s, the introduction of computer-integrated manufacturing (CIM) will change forever the way products are designed, manufactured, and serviced.

With the introduction of the numerous special machines and special cutting tools, production has increased tremendously over standard machine methods. Many products are produced automatically by a continuous flow of finished parts from these special machines. With product control and high production rates, everyone can enjoy the pleasures and conveniences of the automobile, power lawn mowers, automatic washers, stoves, and scores of other products produced today. Without the basic machine tools to produce the first pieces required to develop these ideas of production and automation, the cost of many luxuries that we now enjoy would be prohibitive.

TEST YOUR KNOWLEDGE

1. How has the development of tools helped to improve the standard of living?
2. Name the three general categories of machine tools.
3. What type of work is produced on the following machines?
   (a) Drill press
   (b) Engine lathe
   (c) Metal saw
   (d) Milling machine
   (e) Surface grinder
   (f) Cylindrical grinder
4. What effect has numerical and computer control had on manufacturing?

CHAPTER 3
CAREERS

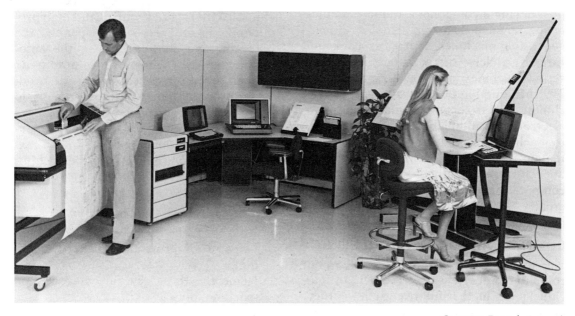

Courtesy Bausch & Lomb

The great advances in technology and the use of computers in the machine tool trade has created many new and specialized positions. To advance in the machine trade, it is necessary to keep up-to-date with modern technology and the ever-widening use of computers in the trade. A young person leaving school may be employed in an average of five jobs during his or her lifetime, three of which do not even exist today. Industry is always on the lookout for bright young people who are precise and who do not hesitate to assume responsibility. To be successful, do your job to the best of your ability and never be satisfied with inferior workmanship. It should be the aim of everyone entering the machine tool trade to produce the best quality product in a reasonable length of time, so that the industry can be competititive with foreign products in world trade.

There are many careers available in the metal working industry. The skill, initiative, and qualifications of an individual will determine which career is most suitable for that person. The machine tool trade will offer exciting opportunities to any ambitious young person who is willing to accept the challenge of working to close tolerances and producing intricate parts.

APPRENTICESHIP TRAINING

Probably the best way to learn any skilled trade is through an apprenticeship program. An apprentice (Fig. 3-1) is one who is employed to learn a trade under the guidance of skilled tradespeople. The apprenticeship program is set up in conjunction with and under the supervision of the company, the Department of Labour, and the trade union. The program is usually about two to four years in duration and includes both on-the-job training and related theory or classroom work. This period of time may be reduced by the completion of approved courses or because of previous experience in the trade.

To qualify for an apprenticeship program, it is advantageous to have completed a high school program or its equivalent. Mechanical ability, with a good standing in mathematics, science, English, and mechanical drawing, is very desirable. Apprentices earn as they learn; the wage scale increases periodically during the training program.

Upon completion of an apprenticeship program a certificate is granted, which qualifies a person to apply as a journeyman in the trade. Further opportunities in the trade are limited only by the initiative and interest of a person. It is quite possible for an apprentice eventually to become an engineer.

MACHINE OPERATOR

Machine operators (Fig. 3-2) are workers who are generally skilled in the operation of one type of machine tool. They are usually rated and paid according to their job classification, skill, and knowledge. The class A operator possesses more skill and knowledge than class B and C operators. For example, a class A operator should be able to operate the machine and
— make necessary machine setups;
— adjust cutting tools;
— calculate cutting speeds and feeds;
— read and understand drawings; and
— read and use precision measuring tools.

Courtesy DoAll Company

Fig. 3-1
An apprentice learns the trade under the guidance of a skilled tradesperson.

Courtesy Cincinnati Milacron Inc.

Fig. 3-2
A machine operator generally operates only one type of machine.

With the continued advancement in tape-controlled machines and programmable robots, there will be fewer operators' jobs. However, machine tool operators who take upgrading technology courses can become operators of computer numerical control (CNC) turning centres, machining centres, and CNC robots.

Courtesy Cincinnati Lathe and Tool Co.

Fig. 3-3
A machinist is skilled in the operation of all machines.

MACHINIST

Machinists (Fig. 3-3) are skilled workers who can efficiently operate all standard machine tools. Machinists must be able to read drawings and use precision measuring instruments and hand tools. They must have acquired enough knowledge and have developed sound judgement to perform any bench, layout, or machine tool operation. In addition, they should be capable of making mathematical calculations required for setting up and machining any part. Machinists should have a thorough knowledge of metallurgy and heat-treating. They should also have a basic understanding of welding, hydraulics, electricity, and pneumatics, and be familiar with computer technology.

Types of Machine Shops
A machinist may qualify to work in a variety of shops. The three most common types are general, production, and jobbing shops.

A *general shop* is usually connected with a manufacturing plant, laboratory, or a foundry. A machinist generally makes and replaces parts for all types of setup and cutting tools and production machinery. The machinist must be able to operate all machine tools and be familiar with bench operations such as layout, fitting, and assembly.

A *production shop* may be connected with a large factory or plant which makes many types of identical machined parts such as pulleys, shafts, bushings, motors, and sheet metal pieces. A person working in a production shop generally operates one type of machine tool and often produces identical parts (Fig. 3-4).

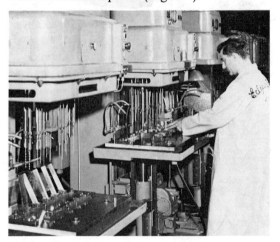

Fig. 3-4
The person in a production shop generally operates one machine and usually produces many identical parts.

A *jobbing shop* is generally equipped with a variety of standard machine tools and perhaps a few production machines such as a turret lathe and punch and shear presses. A jobbing shop may be required to do a variety of tasks, usually under contract to another company. This work may involve the production of jigs, fixtures, dies, moulds, tools, or short runs of special parts. A person working in a jobbing shop generally is a qualified machinist or toolmaker and would be required to operate all types of machine tools and measuring equipment.

TOOL AND DIEMAKER

A *tool and diemaker* is a highly skilled craftsperson who must be able to make different types of dies, moulds, cutting tools, jigs, and fixtures. These tools may be used in the mass production of metal, plastic, or other parts. For example, to make a die to produce a 90° bracket in a punch press, the tool and diemaker must be able to select, machine, and heat treat the steel for the die components. For a mould used to produce a plastic handle in an injection moulding machine, the tool and diemaker must know the type of plastic used, the finish required, and the process used in production.

To qualify as a tool and diemaker, a person should serve an apprenticeship, have above-average mechanical ability, and be able to operate all standard machine tools. This person also requires a broad knowledge of shop mathematics, print reading, machining operations, metallurgy, heat-treating, computers, and space-age machining processes.

TECHNICIAN

A *technician* is a person who works at a level between the professional engineer and the machinist. The technician may assist the engineer in making cost estimates of products, preparing technical reports on plant operation, or programming a numerically controlled machine.

A technician should have completed high school and have at least two years of postsecondary education in a community college, technical institute, or university. A technician must also possess a good knowledge of drafting, mathematics, and technical writing.

Opportunities for technicians are becoming more plentiful because of the development of machine tools such as numerical control, turning centres, and electro-machining processes. Technicians are usually trained in only one area of technology, such as electrical, manufacturing, machine tool, or metallurgy. Technicians specializing in one technology may require a knowledge of more than their specialty field. For example, a machine tool technician (Fig. 3-5) should have a knowledge of industrial machines and manufacturing processes in order to know the best method of manufacturing a product. A technician may qualify as a

Courtesy Bendix Corporation

Fig. 3-5
A technician is often required to check the setup and operation of a machine program.

technologist after at least one year of on-the-job training under a technologist or engineer.

TECHNOLOGIST

A *technologist* works at a level between that of a graduate engineer and a technician. Most technologists are three- or four-year graduates of a community or technical college. Their studies generally include physics, advanced mathematics, chemistry, engineering graphics, computer programming, business organization, and management.

Engineering technologists may do many jobs performed by an engineer, such as design studies, production planning, laboratory experiments, and the supervision of technicians. A technologist may be employed in areas such as quality and cost control, production control, labour relations, training, and product analysis. A technologist may qualify as an engineer by taking further education at university level and by passing a qualifying examination.

PROFESSIONS

There are many areas open to the engineering graduate. Teaching is one of the most satisfying and challenging professions. Graduation from a college in a teacher training course is required, along with on-the-job industrial experience. Although in some cases industrial experience is not a prerequisite, it will prove helpful in teaching. Some provinces and/or schools require on-the-job experience to qualify for technical and vocational certification. With industrial experience, a person may teach industrial arts or in other designated subject areas in a technical or vocational school.

Engineers in industry are responsible for the design and development of new products and production methods and for redesigning and improving existing products. Most engineers specialize in a certain phase of engineering, such as metallurgical, aerospace, mechanical, electrical, and electronics.

A degree in engineering is usually required to enter the profession; however, some classes of engineers, such as tool and manufacturing engineers, progress through a practical experience program and obtain certificates after passing qualifying examinations. Because of the variety of engineering jobs available, women as well as men are entering the many phases of this profession.

Tool and manufacturing engineers are generally responsible for the design and development of a new product. Tool and diemakers can become tool and manufacturing engineers by taking upgrading courses offered at many colleges or by the Society of Manufacturing Engineers. Tool engineers should be able to suggest and design the best method of producing a product quickly and accurately.

NUMERICAL CONTROL AND COMPUTER PROGRAMMERS

Tool programmers for numerically controlled machines must be thoroughly familiar with all machining techniques. A programmer should be familiar with all the sequences and procedures required for producing a machined part. In order to program a machine to produce a part, an operator must be able to
— read working drawings;
— select the best tools for the machining operation;
— calculate speeds and feeds for different materials and types of cutting tools; and

— have a knowledge of production costs and machine tool processes.

To obtain the necessary programming background and knowledge, many vocational schools, technical schools, and universities offer CNC and NC (numerical control) programming courses.

Since the computer is finding such wide use in the metal working trade, a new field is opening up for skilled people who are able to program machine tools to accurately perform any type of machining operation. Computer-aided manufacturing (CAM) is perhaps one of the fastest-growing specialized areas in the machining industry. A good education is essential, along with related information in electronics, mathematics, and manufacturing processes.

TEST YOUR KNOWLEDGE

1. Define an apprentice.
2. Name three desirable qualities a person should have for an apprenticeship training.
3. Explain the difference between a machinist and a machine operator.
4. Briefly explain the difference between a jobbing and a production shop.
5. Define a tool and diemaker.
6. How can a person become a tool and diemaker?
7. Explain the difference between a technician and a technologist.
8. What qualifications are required to become a technical teacher?
9. List four areas in industry that require an engineer's qualifications.

CHAPTER 4
SAFETY

All hand and machine tools can be dangerous if used improperly or carelessly. Working safely is one of the first things a student or apprentice should learn, because the safe way is usually the correct and most efficient way. A person learning to operate machine tools *must first* learn the safety regulations and precautions for each tool or machine. Far too many accidents are caused by carelessness in work habits, or by horseplay. It is easier and much more sensible to develop safe work habits than to suffer the consequences of an accident. SAFETY IS EVERYONE'S BUSINESS AND RESPONSIBILITY.

The safety programs initiated by accident prevention associations, safety councils, governmental agencies, and industrial firms are constantly attempting to reduce the number of accidents. *Nevertheless*, each year accidents which could have been avoided result not only in millions of dollars' worth of lost time and production, but also in a great deal of pain and many lasting physical handicaps. Modern machine tools are equipped with safety features, but it is still the operator's responsibility to use these machines wisely and safely.

CAUSES OF ACCIDENTS

Accidents don't just happen; they are caused. The cause of an accident can usually be traced to carelessness on someone's part. *Accidents can be avoided*, and a person learning the machine shop trade must first develop safe work habits. A safe worker should:

Fig. 4-1
Wearing rings and watches in a machine shop can be the cause of serious injuries.

(a) be neat and tidy at all times;
(b) develop a responsibility to himself or herself;
(c) learn to consider the welfare of fellow workers;
(d) derive satisfaction from performing work accurately and safely.

Good housekeeping is essential to safe working conditions. Good housekeeping is more than cleanliness; it is cleanliness and orderliness. Cultivate the habit of neatness, *have a place for everything, and keep everything in its place.* All working areas should be kept clean and free from obstructions at all times. Grease, oil, tools, and materials left lying around are the main causes of tripping accidents in machine shops. A clean, orderly shop is usually a safe shop in which to work.

Since it would be impossible to list all the causes of accidents in a machine shop, the following general safety suggestions are offered.

1. *Never wear loose clothing around machines. Remove coats and ties and roll up sleeves to the elbows.*
   Loose clothing of any kind may be caught in the moving parts of machinery and draw a person into contact with gears or sharp cutting edges, causing serious injury.

2. *Do not wear rings or watches.*
   These can catch on revolving machine parts and cause severe hand injuries (Fig. 4-1).

3. *Do not operate any machine before understanding its mechanism and knowing how to stop it quickly.*
   Knowing how to stop a machine quickly and practice at doing so can prevent a serious injury.

4. *Keep hands away from moving parts.*
   It is very dangerous practice to "feel" the surface of the revolving work or to stop a machine by hand.

5. *Always stop a machine before measuring, cleaning, or making any adjustments.*
   It is very dangerous to do any type of work around moving parts of a machine.

6. *Never operate a machine unless all safety guards are in place.*
   Safety guards are used to protect an operator from being drawn into moving machine parts (Fig. 4-2).

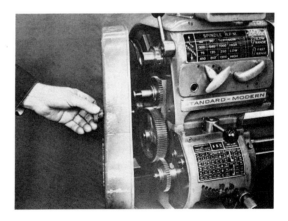

Fig. 4-2
Never operate any machine unless all safety guards are in place.

7. *Always keep the floor around a machine free from oil, grease, tools, and metal cuttings.*

CAUSES OF ACCIDENTS 15

Grease, tools, and parts left on the floor present a tripping hazard (Fig. 4-3). Sharp metal turnings and chips can cut through thin soles or stick to them, making them slippery.

Fig. 4-4
Always wear approved safety glasses to protect your eyes.

Fig. 4-3
Poor housekeeping can lead to accidents.

8. *Never use a rag near the moving parts of a machine.*
   The rag may be drawn into the machine along with the hand that is holding it.
9. *Avoid horseplay at all costs.*
   Too often, shop nonsense results in an accident to an innocent, unsuspecting person.
10. *Always wear safety glasses to protect your eyes* (Fig. 4-4).
    It is good practice and mandatory in many machine shops to wear safety glasses at all times. This is especially important when machining and grinding metal.
11. *Long hair must be protected by a hair net or by an approved shop cap* (Fig. 4-4).
12. *Never have more than one person operate a machine at the same time.*
    Not knowing what the other person would or would not do has caused many accidents.

13. *Get first aid immediately for an injury, no matter how small.*
    Report the injury, and be sure that the smallest cut is treated to prevent the chance of a serious infection.
14. *Always use a brush (not a cloth) to remove chips from a machine* (Fig. 4-5).

WORK AND THINK SAFELY AT ALL TIMES

Fig. 4-5
Always remove chips with a brush or hook; never by hand.

16 CHAPTER 4 / SAFETY

TEST YOUR KNOWLEDGE

1. What is one of the first things a person entering the machine shop trade must learn?
2. List four qualities of a safe worker.
3. Explain how each of the following can contribute to an accident:
   (a) loose clothing
   (b) rings and watches
   (c) poor housekeeping
   (d) horseplay
4. List twenty unsafe acts or conditions found in Fig. 4-6.

Courtesy Industrial Accident Prevention Association of Ontario

Fig. 4-6

CHAPTER 5
MEASURING TOOLS AND SYSTEMS

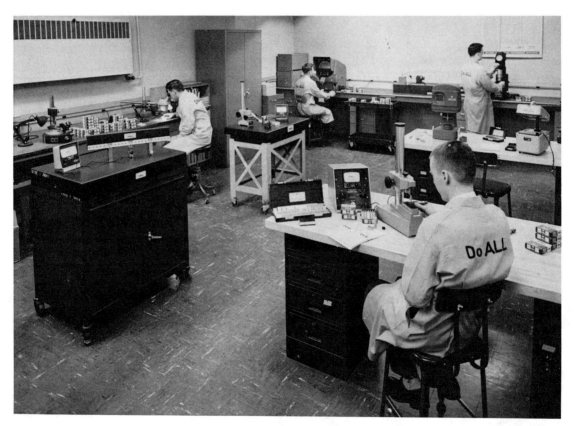

Courtesy DoAll Company

Considerable progress has been made in the science of measurement since the ancient Egyptians used parts of the human body to measure length. Each culture, throughout history, has used one means or another to measure such quantities as length, mass, and time. From such measurements the first standard units of measurement, such as the inch, span, pound, cubit, and second, were developed.

The measuring tools of the past did not have to be too accurate, since most products were custom made by hand. A small margin of error one way or the other made very little difference to the final product. Eli Whitney first conceived the basic idea of mass production through interchangeable parts to fill an order for muskets for the United States government. Interchangeable manufacture was possible only because of improved methods of measurement and mechanically powered machine tools. Through the years, the standards of measurement have been continually improved in order to make them more accurate and reliable. Today's standard of linear measurement is measured in wave lengths of light, bringing the metre to a degree of accuracy within one part in ten million.

The standardization of measurement has made tremendous advances in industrial production and precision possible. Modern automobiles, jet aircraft, and space rockets often require parts finished to an accuracy of less than 0.003 mm (0.0001 in.).

MEASUREMENT SYSTEMS

Currently, two major systems of measurement are used in the world: the *metric (decimal) system*, and the *inch (Imperial and U.S.) system*. Over 90% of the world's population uses some form of the metric system; the inch system is the one that has been traditionally used in Canada and the United States.

METRIC (DECIMAL) SYSTEM

Canada and the United States are currently involved in adopting that version of the metric system known as SI (short for the French, *Système International*). SI is the most advanced and easiest to use system of measurement, and the one which all countries are likely to adopt in time. In SI, the base unit of length is the metre. All other linear units are directly related to the metre by a factor of ten. In order to convert from a smaller to a larger unit, or vice versa, it is necessary to divide or multiply by 10, 100, 1000, etc. For example:

prefix	meaning	multiplier	symbol
micro	one millionth	0.000 001	μ
milli	one thousandth	0.001	m
centi	one hundredth	0.01	c
deci	one tenth	0.1	d
deca	ten	10	da
hecto	one hundred	100	h
kilo	one thousand	1 000	k
mega	one million	1 000 000	M

INCH SYSTEM

The inch (Imperial, U.S.) system has for many years been the standard of measurement for North American industry. In this system the base unit of length is the inch. Other linear units are related to the base unit by odd and unusual factors. The inch can be divided by halves, quarters, eighths, sixteenths, thirty-seconds, sixty-fourths, tenths, hundredths, thousandths, ten-thousandths, etc. Some linear units larger than the inch are:

```
1 foot = 12 inches     =    0.3048 m
1 yard = 36 inches     =    0.9144 m
1 rod  = 198 inches    =    5.0292 m
1 mile = 63 360 inches = 1609.344 m
```

It may readily be seen that the inch system of measurement is far more complex than the SI system. Other quantities, such as weight, volume, pressure, and temperature, are similarly more complex with the inch system than with the SI system. The inch system is even further complicated by the fact that the fluid measurement is different in the Imperial and United States systems.

THE CHANGEOVER PERIOD

Although both Canada and the United States are now committed to conversion to SI as rapidly as possible, it is likely to be some years before all machine tools and measuring devices are redesigned or converted. The change to the metric system in

TABLE 5-1 INCH AND METRIC SYSTEM COMPARISONS				
	Metric		Imperial, United States	
Quantity	Unit	Symbol	Unit	Symbol
Length	metre	m	yard	yd.
	millimetre	mm	inch	in.
Volume	litre	l	quart	qt.
			gallon	gal.
Mass	gram	g	ounce	oz.
(weight)	kilogram	kg	pound	lb.
Force	Newton	N	pound	lb.
Pressure	Pascal	Pa	pounds/square inch	psi
Temperature	degrees Celsius	°C	degrees Fahrenheit	°F
Area	square metre	m²	square feet	sq. ft.

the machine shop trade will be gradual because of the long life expectancy of the costly machine tools and measuring equipment involved. Further, many manufacturers will continue to make inch-sized products as long as their customers request them. It is probable, therefore, that people involved in the machine shop trade will have to be familiar with both the metric and the inch systems during the changeover period.

To accommodate this problem, the following policy has been adopted throughout this book. This policy should enable you to work effectively in both systems now, while permitting an easy transition to full metric as the new materials and tools become available.

(a) Where general measurements or references to quantity are not related specifically to inch standards, tools, or products, only SI units are given.
(b) Where the student may be exposed to equipment designed to both metric and inch standards, separate information is given on both types of equipment in *exact* dimensions.
(c) Where only inch standards, tools, or products exist at the present time, inch measurements are given with a soft conversion to metric provided in parentheses.

STYLE DIFFERENCES BETWEEN SI AND THE INCH SYSTEM

Because SI is an international language, all countries adopting it must follow the approved style. The approved SI style is followed throughout this text wherever standards, tools, products, or processes are described in metric terms. All metric measurements in this text will be followed by the abbreviation for millimetre (mm). If the size is less than one millimetre, the dimension must be preceded by a zero, e.g., 0.97 mm. Inch measurements will be followed by the abbreviation for inch (in.). If the dimension is less than one inch, it is preceded by a zero, e.g., 0.875 in.

CARE OF MEASURING TOOLS

The proper use of the correct measuring tool or instrument plays an important part in the quality and accuracy of the finished product. Precision measuring tools are expensive and should be handled with the greatest of care to preserve their accuracy.

Always keep in mind that an inaccurate measuring tool is worse than no tool at all. Good quality hardened tools will withstand a lifetime of wear with a reasonable amount of care. A good machinist or toolmaker takes pride in keeping measuring tools accurate and in good condition by observing the following rules.
1. Never drop a measuring tool.
2. Never place measuring tools in chips or on oily or dirty surfaces.
3. Always wipe tools clean, and apply a light coating of oil before putting them away.
4. Store measuring tools in separate boxes or cases to avoid accidental scratches and nicks.

STEEL RULES

Steel rules are manufactured with either millimetre or inch graduations. Metric rules are graduated in millimetres and half millimetres. Inch rules are graduated in fractions of an inch. Some rules are available with both millimetre and inch graduations. On these rules, one edge of one side is graduated in thirty-seconds of an inch, with the other edge graduated in millimetres. The other side of the rule has half millimetres on one edge and sixty-fourths of an inch on the other.

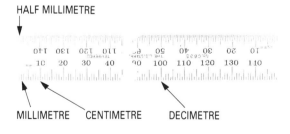

Courtesy L.S. Starrett Co.

Fig. 5-1A
Metric rules are graduated in millimetres and half millimetres.

METRIC RULES

Metric rules (Fig. 5-1A) are generally graduated in millimetres and half millimetres. They are available in lengths from 150 mm to 1 m, the most common being the 150 mm and 300 mm lengths.

INCH RULES

The steel inch rules used in machine shop work are graduated in divisions of 1, 1/2, 1/4, 1/8, 1/16, 1/32, and 1/64 of an inch (Fig. 5-1B). Divisions of 1/64 of an inch are about as fine as can be seen on the inch rule without the use of a magnifying glass. Fractional dimensions given on a print require less accuracy and can be measured with a rule and/or caliper. Inch rules are available in lengths from 1 in. to 72 in. Any dimension given on a print in decimal inches requires the use of precision measuring instruments such as *micrometers and verniers.*

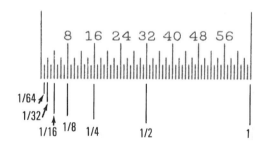

Courtesy L.S. Starrett Co.

Fig. 5-1B
Inch rules are divided into fractions of an inch.

The *spring tempered* (quick-reading) 15 cm or 6 in. rule (Fig. 5-2A) is very common in machine shop work. Metric rules are usually graduated in millimetres and half millimetres.

Inch rules have four separate scales, two on each side. The front is graduated in eighths and sixteenths, and the back is graduated in thirty-seconds and sixty-fourths of an inch. At every eighth of an inch there is a number to make reading in thirty-seconds and sixty-fourths easier and quicker.

Courtesy L.S. Starrett Co.

Fig. 5-2A
A spring tempered (quick-reading) inch rule.

The *flexible* rule is similar to the *spring tempered* rule, but is designed to be more flexible to allow measurements to be made in certain locations where spring tempered rules could not be used.

The *hook rule* (Fig. 5-2B) is used to make accurate measurements from a shoulder, step, or edge of a workpiece. It may also be used to measure flanges, circular pieces, and for setting inside calipers to a dimension.

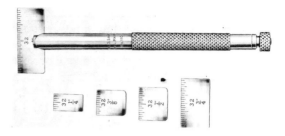

Courtesy L.S. Starrett Co.

Fig. 5-2C
Short length rules are used for measuring small openings.

Short length rules (Fig. 5-2C) are useful in measuring small openings and hard-to-reach locations where an ordinary rule cannot be used. The five small rules in an inch set, the shortest 1/4 in. (6.35 mm) and the longest 1 in. (25.4 mm) long, can be interchanged in the holder.

Measuring Lengths
With a reasonable amount of care, fairly accurate measurements can be made using a steel rule. Whenever possible, butt the end of a rule against a shoulder or step (Fig. 5-3A) to assure an accurate measurement.

Courtesy Brown and Sharpe Manufacturing Company

Fig. 5-2B
A hook rule is used to make accurate measurements from an edge or shoulder.

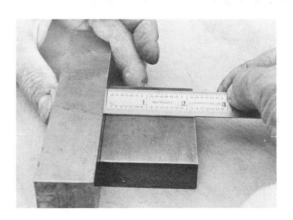

Fig. 5-3A
Butting a rule against a shoulder.

Through constant use, the end of a steel rule becomes worn. Measurements taken from the end are therefore often inaccurate. Accurate measurements of flat work may be made by placing the 1 cm or 1 in. graduation line on the edge of the work, taking the measurement, and subtracting 1 cm or 1 in. from the reading (Fig. 5-3B). When measuring the diameter of round stock it is also advisable to start from the 1 cm or 1 in. graduation line.

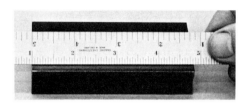

Courtesy Kostel Enterprises Ltd.

Fig. 5-3B
Measuring with a rule, starting at the 1-inch or 1-centimetre line.

The Rule as a Straightedge

The edges of a steel rule are ground flat and may therefore be used as *straightedges* to test the flatness of workpieces. The edge of a rule should be placed on the work surface, which is then held up to the light. Inaccuracies as small as 0.02 mm (0.001 in.) may easily be seen by this method.

OUTSIDE CALIPERS

Outside calipers are tools used to measure the outside surface of either round or flat work. They are made in several styles, such as *spring joint* and *firm joint calipers*. The spring joint caliper (Fig. 5-4) consists of two curved legs, a spring, and an adjusting nut. The outside spring joint caliper is most commonly used because it can easily be adjusted to size. The caliper itself cannot be read directly and therefore must be set to a steel rule or a standard size gauge.

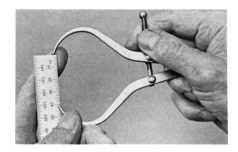

Courtesy Kostel Enterprises Ltd.

Fig. 5-4
Setting an outside caliper to a rule.

Setting the Caliper to a Rule

1. Hold the rule in one hand so that the forefinger extends slightly beyond the end of the rule (Fig. 5-4).
2. Place one leg of the caliper over the end of the rule, and support it with the end of the forefinger.
3. Turn the adjusting nut, with the thumb and forefinger of the other hand, until the end of the "free" caliper leg splits the desired graduation line.
   Keep the caliper legs parallel to the edge of the rule for accurate settings.

Checking the Work Size

Although measuring the work with calipers is not an accurate method of measurement, the work can be reasonably close to size if the proper sense of touch or "feel" is developed by the machinist. Calipers may be used to measure narrow grooves where a micrometer could not be used.
NOTE: Never attempt to measure the work while it is revolving or moving. It is not only a dangerous practice, but the measurement will not be accurate.

1. Hold the caliper spring lightly between the thumb and the forefinger (Fig. 5-5A).
2. Place the caliper on the work so that the line between the tips of the caliper legs is at right angles to the centre line of the work (Fig. 5-5B).
3. Note that the diameter is correct when the caliper just slides over the work with its own weight.

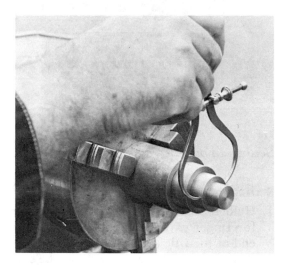

Courtesy South Bend Lathe, Inc.

A

INSIDE CALIPERS

Inside calipers are used to measure the diameter of holes, or the width of keyways and slots. They are made in several styles, such as the *spring joint* and the *firm joint calipers*.

Setting Inside Calipers

Inside calipers, set to a rule, may be used to measure work to a rough size only. When checking the finished size of a workpiece, the calipers should be set to a gauge or a micrometer.

1. Butt the end of a rule and one leg of the caliper against a shoulder or flat surface (Fig. 5-6).
2. Keep the ends of both legs parallel to the edge of the rule.
3. With the thumb and forefinger, turn the adjusting nut until the tip of the free leg splits the desired graduation.

Measuring Inside Diameters

1. Place one leg of the caliper at the bottom of the hole. Keep the leg from moving by holding it in place with the forefinger of one hand (Fig. 5-7).

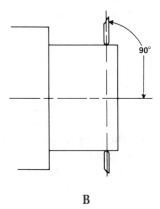

B

Fig. 5-5
Checking a diameter with an outside caliper.

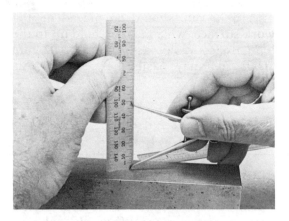

Courtesy Kostel Enterprises Ltd.

Fig. 5-6
Setting an inside caliper to a rule.

2. Adjust the caliper with the thumb and forefinger of the other hand until it cannot be moved sideways and a slight drag is felt on the free leg as it is moved in and out of the hole (Fig. 5-7).

Transferring Measurements

Inside calipers which have been set to a hole may be checked for approximate size with a rule (Fig. 5-6).

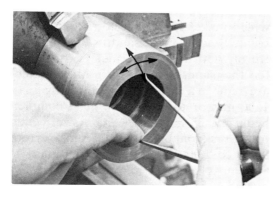

Courtesy Kostel Enterprises Ltd.

Fig. 5-7
Adjusting an inside caliper to the size of a hole.

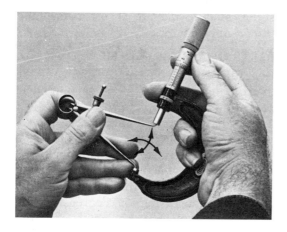

Courtesy Kostel Enterprises Ltd.

Fig. 5-8
Transferring an inside caliper setting to a micrometer.

When an accurate measurement is required, however, the caliper setting should be tested with an outside micrometer.

1. Hold the micrometer in one hand so that it can be adjusted with the thumb and forefinger (Fig. 5-8).
2. Place one leg of the caliper on the micrometer anvil and hold it in position with one finger.
3. Move the other leg across the micrometer spindle and adjust the micrometer until a slight drag is felt.

   The amount of drag on the caliper leg must be the same as the drag felt in the hole.
4. Remove the thumb and forefinger from the micrometer and note the reading.

DECIMAL INCH SYSTEM

When inch units of measurement smaller than a sixty-fourth of an inch are required, the decimal inch system is used. The inch is divided into ten equal parts, each having a value of one hundred thousandths (0.100). Each of these is again divided into ten equal parts and so on until the inch is divided into tenths, hundredths, thousandths, and ten-thousandths of an inch (Fig. 5-9).

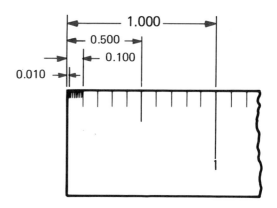

Fig. 5-9
A 1-inch space divided into tenths and hundredths of an inch, using the decimal system.

The most commonly used decimal inch fraction is one thousandth of an inch or 0.001 in. (three numbers to the right of the decimal point). Figures to the left of the decimal point are whole numbers representing whole inches. The figures to the right of the decimal represent an amount that is less than one inch. For example, a dimension of 2.725 in. means two inches plus seven hundred and twenty-five thousandths of an inch. If the dimension is less than one inch, a zero is placed to the left of the decimal point. For example, 0.825 in. means a measurement of eight hundred and twenty-five thousandths of an inch.

Sometimes dimensions are given with only one or two numbers to the right of the decimal point. In such cases add enough *zeros* (0) to have three numbers to the right of the decimal: e.g., for the dimension 0.15 in. add one zero (0) to make it 0.150 in. (one hundred and fifty thousandths of an inch); for the dimension 0.5 in. add two zeros (00) to make it 0.500 in. (five hundred thousandths of an inch).

> The addition of zeros to the right of the last whole number does not change the value of the dimension.

Various decimal fractions and their meanings are listed below:

0.001 means one thousandth
0.010 means ten thousandths (sometimes written 0.01)
0.025 means twenty-five thousandths
0.100 means one hundred thousandths (sometimes written 0.1)
0.325 means three hundred and twenty-five thousandths
0.750 means seven hundred and fifty thousandths (sometimes written 0.75)

The SI system is already based on the decimal system. When it is necessary to measure lengths smaller than 0.5 mm, a micrometer should be used. Metric micrometers divide each millimetre into one hundred equal parts, providing readings in hundredths of a millimetre (0.01 mm).

MICROMETER CALIPERS

Micrometers are precision measuring tools used for making accurate measurements. Both metric and inch standard micrometers are available, graduated to read either in hundredths of a millimetre or in thousandths of an inch. However, some have a vernier scale on the barrel, which makes it possible to take measurements to within thousandths of a millimetre or, with inch micrometers, to within ten thousandths of an inch. Micrometers are made in a variety of sizes ranging up to 1500 mm or to 60 in., the most common being the 25 mm or the 1 in. size, which measures from 0.00 to 25.00 mm, or from 0.000 to 1.000 in. for inch micrometers. The most common micrometers used in machine shop work have a 25 mm or a 1 in. range; for example, a 150 mm micrometer would measure from 125 to 150 mm. A 2 in. micrometer will measure from 1.000 to 2.000 in. The range of special purpose micrometers may vary. Micrometers are available in a wide variety of types and styles, such as *outside, inside, depth,* and *thread* micrometers.

PARTS OF A MICROMETER

Regardless of the type or size of an outside micrometer, they all contain certain basic parts (Fig. 5-10).

The U-shaped *frame* holds all the micrometer parts together.

The *anvil* is the fixed measuring face.

The *spindle* is the moveable measuring face. Turning the thimble moves the spindle, increasing or decreasing the distance between the anvil and the spindle face.

The *sleeve* (barrel) is graduated into equal divisions, each line having a value of 0.5 mm for metric micrometers or 0.025 in. for inch micrometers.

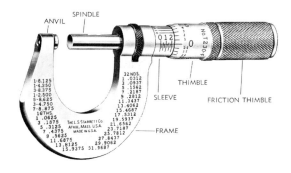

Courtesy L.S. Starrett Co.

Fig. 5-10
The main parts of an outside micrometer.

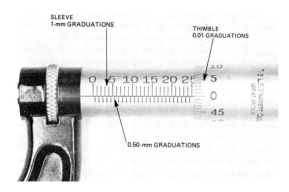

Courtesy Kostel Enterprises Ltd.

Fig. 5-11
The graduations on a metric micrometer.

The *thimble* has equally spaced divisions around its circumference, each having a value of 0.01 mm for metric micrometers or 0.001 in. for inch micrometers.

The *friction thimble* on the end of the thimble is used to assure an accurate measurement, and to prevent too much pressure being applied to the micrometer. On some micrometers, a *ratchet stop* serves the same purpose.

DESCRIPTION OF A METRIC MICROMETER

The pitch of the micrometer screw on metric micrometers is 0.50 mm. Therefore, one complete revolution of the spindle will either increase or decrease the distance between the measuring faces 0.50 mm. The graduations on the *sleeve* (barrel) above the index line, are in whole millimetres (from 0 to 25). The graduations below the index line represent 0.50 mm, the pitch of the micrometer thread (Fig. 5-11). Since a complete turn of the micrometer spindle will move the measuring faces 0.50 mm, two complete turns are required to move the spindle 1 mm.

The circumference of the *thimble* is graduated into 50 equal divisions, with every fifth line being numbered. Since one revolution of the thimble advances the spindle 0.50 mm, each graduation on the thimble equals 1/50 × 0.50 or 0.01 mm.

To Read a Metric Micrometer

1. Note the last number showing on the sleeve; multiply this by 1 mm.
2. Note the number of lines (above and below the index line) which show past the numbered line; multiply these by 0.50 mm.
3. Add the number of the line on the thimble that coincides with the index line.

In Fig. 5-12:

#10 is showing on the sleeve	10 × 1 = 10.00
5 lines past the #10 on the sleeve	5 × 0.50 = 2.50
Total reading	12.50 mm

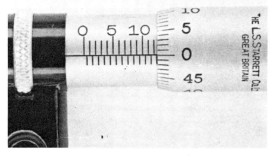

Courtesy Kostel Enterprises Ltd.

Fig. 5-12
A metric micrometer showing a reading of 12.50 mm.

DESCRIPTION OF A STANDARD INCH MICROMETER

There are 40 threads per inch on the micrometer spindle. Therefore, one complete revolution of the spindle either increases or decreases the distance between the measuring faces 1/40 of an inch, or 0.025 in.

The one-inch distance, marked on the micrometer sleeve (Fig. 5-13), is divided into 40 equal divisions, each of which equals 1/40 (0.025) in. If the micrometer is closed completely until the zero mark on the thimble is aligned with the centre (index) line on the barrel, and then the thimble is revolved counterclockwise one complete revolution, it will be noted that a line has appeared on the sleeve (barrel). Each line on the sleeve indicates 0.025 in. Thus, if 3 lines are showing on the barrel, the micrometer is opened 3 × 0.025 in. or 0.075 in.

Every *fourth* line is longer than the others, and is numbered. Each numbered line indicates a distance of 0.100 in. For example, a number 4 showing on the sleeve indicates a distance between the measuring faces of 4 × 0.100 or 0.400 in.

The thimble has 25 equal divisions about its circumference, each of these divisions equalling 0.001 in.

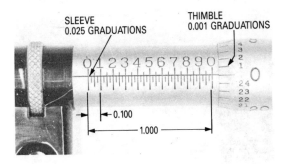

Courtesy L.S. Starrett Co.

Fig. 5-13
The graduations on a standard inch micrometer.

To Read An Inch Micrometer

To read an inch micrometer, note the last number showing on the sleeve, multiply this by 0.100, multiply the number of small lines visible past that number by 0.025, and add the number of divisions on the thimble from zero to the line that coincides with the centre or index line on the sleeve.

In Fig. 5-14:

#2 is shown on the sleeve	2 × 0.100 = 0.200
1 line is visible past #2	1 × 0.025 = 0.025
16 divisions past the zero on the thimble	16 × 0.001 = 0.016
Total reading	0.241 in.

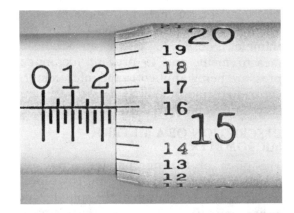

Courtesy L.S. Starrett Co.

Fig. 5-14
An inch micrometer showing a reading of 0.241 in.

CARE OF THE MICROMETER

Everyone should be impressed with the fact that any instrument capable of measuring to within 0.01 mm (0.0001 in.) must be treated with a great deal of care and not abused; otherwise, its accuracy will be damaged. The following points should be kept in mind when using micrometers:

1. Never drop a micrometer; always place it down gently in a clean place.
2. Never place tools or other materials on a micrometer.
3. Never lay a micrometer in steel chips or grinding dust, or handle it with oily hands.
4. Never attempt to use a micrometer on moving work.
5. Keep the micrometer clean and accurately adjusted.

TESTING THE ACCURACY OF MICROMETERS

The accuracy of a micrometer should be tested periodically to assure that the work produced is the size required. *Always* make sure that both measuring faces are clean before a micrometer is checked for accuracy.

To test a micrometer, first clean the measuring faces and then turn the thimble, using the friction thimble or ratchet stop, until the measuring faces contact each other. If the zero line on the thimble coincides with the centre (index) line on the sleeve, the micrometer is accurate. Micrometers can also be checked for accuracy by measuring a gauge block or known standard (Fig. 5-15). The reading of the micrometer must be the same as the gauge block or standard. Any micrometer which is not accurate should be adjusted by a qualified person.

MEASURING WITH A MICROMETER

When measuring with a micrometer, make sure that the measuring faces are clean and that the micrometer is held squarely across the work. The thimble should never be adjusted too tightly, or the micrometer may be damaged. The correct tension or *feel* may be checked by closing the spindle on a hair or piece of paper until it pulls snugly between the anvil and spindle. It is wise to use the friction thimble or ratchet stop whenever possible, in order to get the correct tension and measurement.

Figure 5-16 shows how a micrometer should be held in order to accurately measure a piece of work held in the hand.

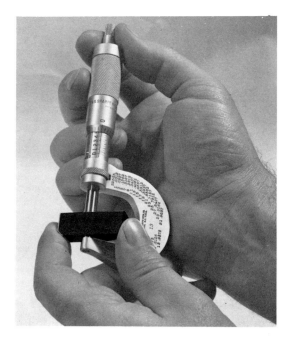

Courtesy L.S. Starrett Co.

Fig. 5-15
Checking the accuracy of a micrometer using a gauge block.

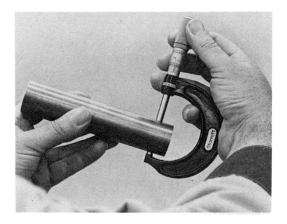

Courtesy Kostel Enterprises Ltd.

Fig. 5-16
The correct way to hold a micrometer when measuring work held in the hand.

Carefully note the position of the fingers. The micrometer is held against the palm of the hand by the little or third finger, and the thumb and forefinger are used to turn the thimble.

When measuring with a micrometer, move it slightly to the left and right while turning the thimble. This helps to align the micrometer correctly on the workpiece.

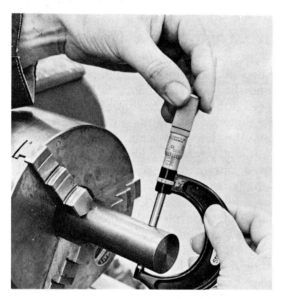

Courtesy Kostel Enterprises Ltd.

Fig. 5-17
The correct way to hold a micrometer when measuring work in a machine.

The proper way of holding a micrometer when measuring work in a machine is illustrated in Fig. 5-17. Hold the top of the micrometer frame in the right hand, place the micrometer over the work, and turn the thimble with the thumb and forefinger of the left hand.

VERNIER CALIPERS

The vernier caliper (Fig. 5-18) is a precision measuring instrument used to make accurate measurements to within 0.02 mm, or to within 0.001 in. for inch tools. The bar and movable jaw are graduated on both sides, one side for taking outside measurements and the other side for inside measurements. Vernier calipers are available in metric and in inch graduations, and some types have both scales (Fig. 5-19). The parts of the vernier caliper are the same, regardless of the measurement system for which the instrument is designed.

Fig. 5-18

Fig. 5-19
A vernier caliper with metric and inch graduations.

DESCRIPTION OF METRIC VERNIER CALIPERS

The *main scale* on the bar is graduated in millimetres, and every main division is numbered. Each numbered division has a value of 10 millimetres; for example, #1 represents 10 mm, #2 represents 20 mm, etc. There are 50 graduations on the sliding or *vernier scale*, with every fifth one being numbered. These 50 graduations occupy the same space as 49 graduations on the main scale (49 mm).

Therefore 1 vernier division
$$= \frac{49}{50}$$
$$= 0.98 \text{ mm}$$

The difference between 1 main scale division and 1 vernier scale division
$$= 1 - 0.98$$
$$= 0.02 \text{ mm}$$

To Read a Metric Vernier Caliper

1. The last numbered division on the bar to the left of the vernier scale represents the number of millimetres multiplied by 10.
2. Note how many full graduations are showing between this numbered division and the zero on the vernier scale. Multiply this number by 1 mm.
3. Find the line on the vernier scale which coincides with a line on the bar. Multiply this number by 0.02 mm.

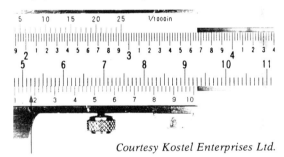

Courtesy Kostel Enterprises Ltd.

Fig. 5-20
A metric vernier caliper reading of 43.18 mm.

In Fig. 5-20:
The large #4 graduation on the bar	(4 × 10 mm)	= 40
Three full lines past the #4 graduation	(3 × 1 mm)	= 3
The 9th line on the vernier scale coincides with a line on the bar	(9 × 0.02)	= 0.18
Total reading		= 43.18 mm

DESCRIPTION OF INCH VERNIER CALIPERS

Two types of vernier inch calipers are available: one has 25 divisions on the vernier scale of the movable jaw, and the other type has 50 divisions. The 50-division scale is much easier to read than the 25-division scale.

The 25-Division Vernier Inch Caliper

The bar is graduated in exactly the same way as the sleeve of an inch micrometer. Each inch is divided into 40 equal divisions, each having a value of 0.025 in. Every fourth line representing 1/10 in. or 0.100 in. is numbered. The vernier scale on the movable jaw has 25 equal divisions, each having a value of 0.001 in. The 25 divisions on the vernier scale occupy the same space as 24 divisions on the bar. Therefore, only one line of the vernier scale will line up exactly with a line on the bar at any one setting.

To read a 25-division vernier, note how many inches, hundred thousandths (0.100) and twenty-five thousandths (0.025) the zero mark on the movable jaw is past the zero mark on the bar. To this total, add the number of thousandths (0.001) indicated by a line on the vernier scale which exactly coincides with a line on the bar.

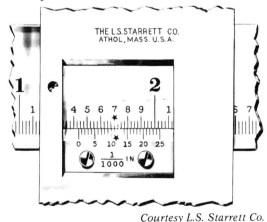

Courtesy L.S. Starrett Co.

Fig. 5-21
A 25-division vernier caliper reading of 1.436 in.

In Fig. 5-21:
the large #1 on the bar = 1.000
the small #4 past
 the #1 (4 × 0.100) = 0.400
1 line visible past
 the #4 (1 × 0.025) = 0.025
the 11th line of the
 vernier coincides
 with a line
 on the bar (11 × 0.001) = 0.011
Total reading = 1.436 in.

The 50-Division Vernier Inch Caliper

Each line on the bar of the vernier represents 0.050 in. Every second line is numbered and represents 0.100 in. The vernier scale on the movable jaw had 50 equal divisions, each having a value of 0.001 in. The fifty divisions on the vernier scale occupy the same space as forty-nine divisions on the bar. Therefore only one line of the vernier scale will line up exactly with a line on the bar at any one setting.

To read a 50-division vernier, note how many inches, hundred thousandths (0.100), and fifty thousandths (0.050) the zero mark on the movable jaw is past the zero mark on the bar. To this total add the number of thousandths (0.001) indicated by the line on the vernier scale which exactly coincides with a line on the bar.

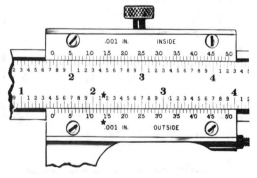

Courtesy L.S. Starrett Co.

Fig. 5-22
A 50-division vernier caliper reading of 1.464 in.

In Fig. 5-22:
The large #1 on the bar = 1.000
The small #4 past
 the #1 (4 × 0.100) = 0.400
1 line visible past
 the #4 (1 × 0.050) = 0.050
The 14th line
 on the vernier scale
 coincides with a
 line on the
 bar (14 × 0.001) = 0.014
Total reading 1.464 in.

TEST YOUR KNOWLEDGE

1. Who first conceived the basic idea of mass production through interchangeable parts?

Systems of Measurement
2. What is the common unit of length in (a) the inch system? (b) SI?
3. Name several advantages that SI has over the inch system.

Care of Tools
4. State two reasons why measuring tools should be handled with care.
5. What should be done before storing measuring tools?
6. Why should tools be stored in separate boxes or cases?

Steel Rules
7. What is the smallest graduation which can be clearly seen on a metric rule? On an inch rule?
8. Name two types of measuring tools that can be used when a reading on a print is given in decimals.
9. Name four types of steel rules used in machine shop work.
10. What is the most common type of rule used in machine shop work?
11. State the purpose of:
 (a) hook rules
 (b) short-length rules

Measuring Lengths
12. Explain how accurate measurements

can be made if the end of the rule is worn.
13. How can a rule be used as a straightedge?

Outside Calipers
14. Name two types of outside calipers.
15. Briefly explain how to set an outside caliper to size.
16. Why is it dangerous to try to measure work while it is revolving?
17. Explain how an outside caliper should be held when measuring work.

Inside Calipers
18. For what purpose are inside calipers used?
19. Explain how an inside caliper may be set to a size of 31.75 mm (1-1/4 in.).
20. Describe how inside calipers should be held and adjusted when measuring an inside diameter.
21. Name two methods of transferring inside caliper measurements.

Decimal Inch System
22. Why are decimal fractions necessary?
23. What common decimal inch fraction is sometimes used in machine shop work?
24. What do the figures to the right and to the left of the decimal point represent?
25. Explain what should be done when there are only one or two numbers to the right of the decimal point.

Micrometer Calipers
26. What is the range of the most common metric and inch micrometers?
27. Name and state the purpose of four main parts of a micrometer.
28. What is the pitch of a metric micrometer screw thread?
29. For metric micrometers what is the value of:
    (a) each line on the sleeve?
    (b) each numbered line on the sleeve?
    (c) each line on the thimble?
30. What is the reading of the following metric micrometer settings?

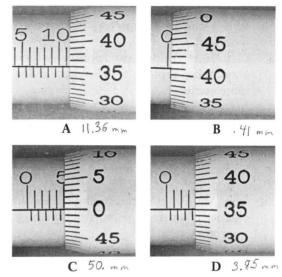

A 11.36 mm B .41 mm
C 50. mm D 3.95 mm

Courtesy Kostel Enterprises Ltd.

Fig. 5-23

31. How many threads per inch are there on a standard inch micrometer?
32. What is the value of the following inch micrometer graduations?
    (a) each line on the sleeve
    (b) each line on the thimble
33. What is the reading of the following inch micrometer settings?

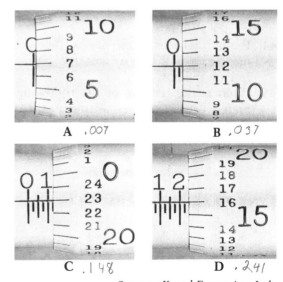

A .007 B .037
C .148 D .241

Courtesy Kostel Enterprises Ltd.

Fig. 5-24

Care of the Micrometer
34. Why should micrometers be handled with care?
35. List four important points which should be kept in mind when using micrometers.

Testing the Accuracy of Micrometers
36. Why should a micrometer be tested for accuracy?
37. Describe two methods of testing the accuracy of a micrometer.

Measuring with a Micrometer
38. Why should the measuring faces be cleaned before measuring with a micrometer?
39. Describe how the correct tension or feel on a micrometer can be checked.
40. Explain how to hold a micrometer when measuring work held in the hand.
41. Explain how to hold a micrometer when measuring work in a machine.

Vernier Calipers
42. How is a vernier caliper graduated in order to make outside and inside measurements?
43. How are the following parts of a 25-division vernier inch caliper graduated?
    (a) the bar
    (b) the sliding jaw
44. Explain why only one line of the vernier scale will line up exactly with a line on the bar.
45. What is the reading of the following metric vernier caliper settings?

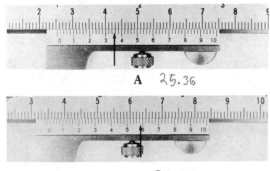

A 25.36

B 34.58

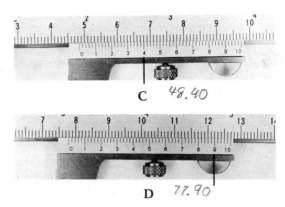

C 48.40

D 77.90

Fig. 5-25   Courtesy Kostel Enterprises Ltd.

46. What is the value of each line on the sliding jaw of the metric vernier calipers?
47. What are the following inch vernier caliper settings?

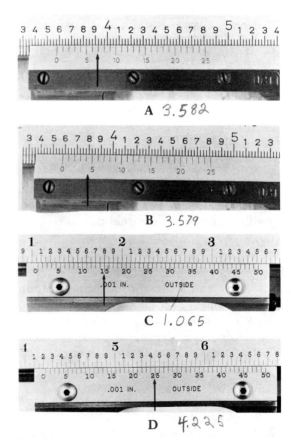

A 3.582

B 3.579

C 1.065

D 4.225

Fig. 5-26   Courtesy Kostel Enterprises Ltd.

CHAPTER 6
LAYOUT TOOLS

A very important function of any machinist is to be able to do accurate layout work. Before any work is machined or holes are drilled, it is necessary first to lay out the amount of material that is to be removed or to locate the position of the holes. The dimensions or specifications, which represent the actual size of the finished workpiece, are usually found on a print prepared by a drafter or tool designer. A competent machinist must be able to read and understand drawings or prints, select and use the proper layout tools, and transfer the dimensions from the print to the metal workpiece.

Laying out is the process of scribing (marking) centre points, circles, arcs, or straight lines on metal to indicate the position of holes to be drilled or the amount of material to be removed on a shaper or milling machine. All layouts should be made from a *baseline* or a machined edge to ensure the accuracy of the layout and the correct position of dimensions in relation to each other. The accuracy of a finished job is generally determined by the accuracy and amount of care used while laying out.

PREPARING SURFACES FOR LAYOUT

Layout lines must be plain to see. Therefore the surface of the work must first be coated with some type of layout material, so that scribed lines will stand out in sharp contrast to the background of the layout material. Before any type of layout material is applied to work, the surface must be clean and free from surface scale, grease, or oil; otherwise the layout material will not stick to it. There are numerous ways of coating surfaces for layout, and a few are listed below:

1. A commercial layout dye or bluing is generally used to coat work surfaces. It is inexpensive and fast drying. Layout dye is generally applied with a brush, or sprayed on with an aerosol can.
2. Chalk may be rubbed into the rough surface of castings.
3. A copper sulphate solution, called blue vitriol, can be used for coating machined surfaces.
   NOTE: Apply blue vitriol only to ferrous metals.
4. Certain metals may be heated to a blue colour. When the metal is cooled, scribed layout lines may be clearly seen against the blue background.

SURFACE PLATES

Surface plates (Fig. 6-1) are used where fine accuracy is required in layout work. Surface plates are made from a good grade of cast iron, granite, or ceramic material. Because of their flat surface, surface plates provide a reference or starting point for layout operations. Some granite surface plates are lapped flat to within

Courtesy Taft-Peirce Manufacturing Co.

Fig. 6-1
A surface plate provides a true, flat reference surface when laying out work.

0.0025 mm (0.0001 in.). Besides being used for accurate layout work, they are also used for inspecting gauges, jigs, and fixtures. To maintain the accuracy of a surface plate and the layout, the top should be kept clean and free of burrs.

SCRIBER

The scriber is a tool used for marking layout lines on surfaces. Scribers are made of tool steel, about 5 mm (3/16 in.) in diameter, with hardened and tempered points. In order for a scriber to mark fine, clear layout lines, it is important that its point be sharp. Never perform any layout work with a dull scriber. Two of the most common types of scribers are illustrated in Fig. 6-2.

Courtesy Eclipse-Pioneer

Fig. 6-2
A pocket and a double-end scriber.

Scribing a Line

Straight lines may be scribed on metal surfaces by *drawing* the scriber along the edge of a square or rule. Incline the scriber at a slight angle to keep the point tight against the edge of the square or rule.

Since indistinct or double lines are useless, always make sure the scriber point is sharp.

PRICK PUNCH

A prick punch (Fig. 6-3) is a layout instrument made of tool steel, about 100 to 150 mm (4 to 6 in.) long, with both ends hardened and tempered. Its point is ground to an angle of from 30 to 60°. It is used to make small indentations along layout lines, to mark centres for drilled holes, and also centres for divider points. Punch marks are sometimes called *witness* marks, because the indentations will still remain if the layout lines should be rubbed off the surface of the work.

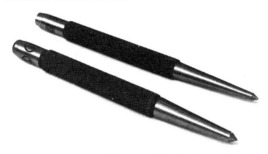

Courtesy Kostel Enterprises Ltd.

Fig. 6-3
A prick punch and a centre punch are used in layout work.

CENTRE PUNCH

The centre punch (Fig. 6-3) is similar to a prick punch, but its point is ground to an angle of approximately 90°. A centre punch is used to enlarge the prick punch marks so that a drill may be started easily and accurately. Some centre punches are automatic, with the striking mechanism enclosed in the handle (Fig. 6-4). A downward pressure on the handle releases the striking mechanism and makes the impression.

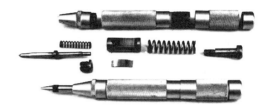

Courtesy Kostel Enterprises Ltd.

Fig. 6-4
An automatic centre punch contains a striking mechanism in the handle.

The following points should be kept in mind while using either a centre or prick punch:
1. Always make sure the point of the punch is *sharp*.
2. Hold the punch at a 45° angle and place the point on the layout line.
3. Bring the punch to a vertical position and tap it gently with a light hammer.
4. Examine the position of the punch mark and correct it if necessary.

DIVIDER

The divider (Fig. 6-5) is a tool with hardened steel points used for transferring measurements, comparing distances, and

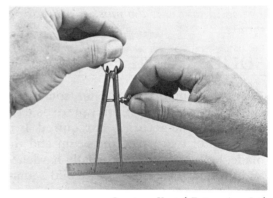

Courtesy Kostel Enterprises Ltd.

Fig. 6-5
Setting a divider using the 1 in. or 1 cm line.

scribing arcs and circles. Dividers are adjustable and are classified according to size by the maximum opening between the two points.

To set the divider to a size, place one point in the 1 cm or the 1 in. line of steel rule and adjust the other point until it splits the graduation line the correct distance away. The graduation lines on a rule are V-shaped, and often a divider may be set more exactly by feeling than by seeing. If an extremely accurate size is needed, check the setting with a magnifying glass.

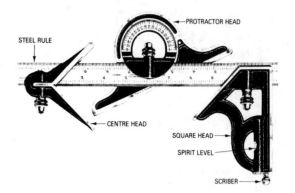

Courtesy L.S. Starrett Co.

Fig. 6-7
The main parts of a combination set.

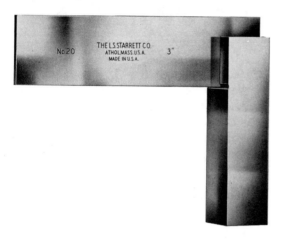

Courtesy L.S. Starrett Co.

Fig. 6-6
The solid square is the most accurate square.

SOLID SQUARE

The solid square (Fig. 6-6), sometimes called the *master precision* square, is used where extreme accuracy is required. It is made up of two parts, the beam and the blade, both of which are usually hardened and ground. Three uses of a solid square are:
(a) to check a surface for flatness,
(b) to determine if two surfaces are at right angles (square) to each other,
(c) to check the accuracy of other squares.

COMBINATION SET

The combination set (Fig. 6-7) is one of the most useful and versatile tools in a machine shop. The set consists of four principal parts: steel rule, square head, bevel protractor, and centre head.

STEEL RULE

The steel rule or blade may be fitted to the centre head, the bevel protractor, or to the square head. Sometimes it is used separately as a straightedge or for measuring. Metric combination set rules are usually graduated in millimetres and half millimetres. Inch combination set rules are usually graduated in eighths and sixteenths on one side and in thirty-seconds and sixty-fourths on the other side.

SQUARE HEAD

The square head or combination square is used to lay out lines parallel and at right angles to an edge. It may also be used as a depth gauge or for checking 45° and 90° angles. The square head can be moved along to any position on the rule.

To Lay Out Parallel Lines (Fig. 6-8A)
1. If possible, hold the work in a vise to prevent it from moving during the layout operation.

2. Remove all burrs from the edge of the work with a file.
3. Extend the steel rule the desired distance beyond the body of the square. Always be sure that only *half the graduation line* can be seen.
4. Hold the body of the square tightly against a *machined edge* with the rule flat on the work surface.
5. Hold the scriber at a *slight angle* (Fig. 6-8A) to keep the point against the end of the rule.

Courtesy Kostel Enterprises Ltd.

Fig. 6-8A
Using the combination square to lay out a line parallel to a machined edge.

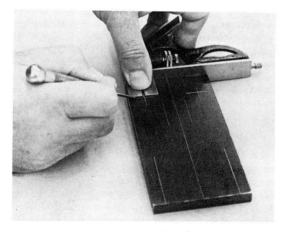

Courtesy Kostel Enterprises Ltd.

Fig. 6-8B
Laying out lines at right angles.

6. Draw a sharp line along the end of the rule.
7. Move the square a short distance and again scribe along the end of the rule.
8. Continue moving the square and scribing until the line is complete.

Lines at *right angles*, or the location of holes, may be laid out by placing the body of the square against a machined edge of the work that is at right angles (90°) to the first edge (Fig. 6-8B). Follow the same procedure for laying out lines at right angles as was outlined for laying out parallel lines.

BEVEL PROTRACTOR

The bevel protractor is used to lay out and check angles. The protractor can be adjusted from 0 to 180°. On some protractors, the scale is graduated from 0 to 90° from both right and left.

CENTRE HEAD

The centre head forms a centre square when clamped to the rule. It can be used for locating centres of round, square, and octagonal stock.

This tool can be used effectively only when the stock being centred is true in shape (round, square, octagonal) and the ends have been machined square.

HERMAPHRODITE CALIPER

The hermaphrodite caliper (Fig. 6-9) is generally used to locate the centres of round work or work which has been cast and is not quite round. It has one bent leg and one straight leg which contains a sharp point used to scribe layout lines. Hermaphrodite calipers may also be used to scribe lines parallel with a machined edge or shoulder (Fig. 6-10). When setting this tool to a size, place the bent leg on the edge of a rule and adjust the other leg until the scriber point is at the desired graduation.

Fig. 6-9
A hermaphrodite caliper may be used to locate the centre of a round stock.

CENTRING WORK

Work that is to be machined between centres on a lathe must have a centre hole drilled in each end to support the workpiece in the lathe centres. Although these centre holes are more accurately and easily drilled on a lathe with the work being held in a three-jaw chuck, other methods may be used to centre the stock. Some of the more common methods used to lay out the centres of round stock are by the use of the centre head, the hermaphrodite caliper, or the bell centre punch.

To Locate the Centres of Round Stock

(a) *Centre Head Method* (Fig. 6-11)

The centre head offers a quick and reasonably accurate method of centre location.

1. Place the work in a vise and remove the burrs or sharp edges with a file.
2. Apply layout dye to both ends of the work.
3. Hold the centre head firmly against the work with the rule flat on the end.
4. Hold a *sharp scriber* at an angle so that its point touches the edge of the rule.

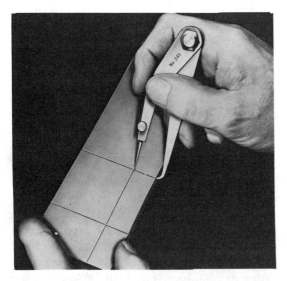

Courtesy L.S. Starrett Co.

Fig. 6-10
Scribing a line parallel to an edge with a hermaphrodite caliper.

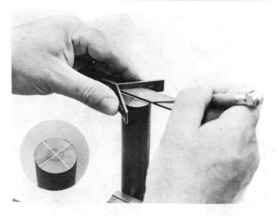

Courtesy Kostel Enterprises Ltd.

Fig. 6-11
Using a centre head to locate the centre of round stock.

40   CHAPTER 6 / LAYOUT TOOLS

5. Scribe a line along the edge of the rule.
6. Rotate the centre head one-quarter of a turn and scribe a second line.
7. *Lightly* centre punch where the two lines cross.
8. Repeat steps 3 to 7 on the other end of the work.
9. Test the accuracy of the centre layout.

(b) *Hermaphrodite Caliper Method*
   The hermaphrodite caliper is often used to locate centres when a centre head cannot be used.
   1. Place the work in a vise and remove the burrs or sharp edges with a file.
   2. Apply layout dye to both ends.
   3. Set the hermaphrodite caliper to approximately one-half the work diameter.
   4. With the thumb of one hand, hold the bent leg just below the edge of the work and scribe an arc (Fig. 6-9).
   5. Move the bent leg a quarter of a turn and scribe an arc. Repeat until four arcs are scribed on each end of the work.
   6. Lightly centre punch the centre of the four arcs and check the accuracy of the centres.

(c) *Bell Centre Punch Method*
   The bell centre punch (Fig. 6-12A) is a quick and reasonably accurate method of locating the centres on a piece of round stock. It consists of a bell-shaped housing and a round, pointed rod which is free to slide in the centre of the housing.
   1. Face both ends of the workpiece and remove the burrs so that the bell will sit squarely on the work.
   2. Place the work in a vise.
   3. Place the bell centre punch over the work (Fig. 6-12B) and hold it vertical.
   4. Sharply strike the punch with a hammer.

A

B

Fig. 6-12
A bell centre punch.

To Check the Accuracy of the Centre Layout

Before any metal is cut or a hole is drilled, it is good practice to check the accuracy of the layout. This is especially true when laying out lathe centre holes, since an error in layout will cause the centre hole to be drilled off centre. The easiest method used to check the centre hole layout is to use a pair of dividers.

Divider Method

1. Place one leg of the divider in the light centre punch mark.
2. Adjust the divider so that the other leg is on a line and set exactly to the edge of the work (Fig. 6-13).
3. Revolve the divider one-half turn (to the other end of the same line) and check if the leg of the divider is in the same relation to the edge of the work as in step 2.
4. If the divider leg is not in the same relation to the work edge at both ends of the line, lightly move the centre punch mark to correct the error (Fig. 6-14).
5. Repeat step 4 on the other scribed line until the punch mark is exactly on centre.
6. Deepen the centre punch mark in preparation for drilling the centre hole.

Courtesy Kostel Enterprises Ltd.

Fig. 6-14
Moving the centre punch mark to bring it to centre.

SURFACE GAUGE

The surface gauge is an instrument used on a surface plate or any flat surface for scribing lines in layout work. It consists of a heavy base and an upright spindle to which a scriber is clamped. The base of the surface gauge has a V groove which allows it to be used on cylindrical work as well as on flat surfaces. There are also pins in the base, which may be pushed down so that the surface gauge can be used against the edge of a surface plate or a slot. A surface gauge can also be used as a height gauge and for levelling work in a machine vise.

A surface gauge can be used to scribe a series of parallel lines on any workpiece set on the top of a surface plate. If a piece of work is fastened to an angle plate and mounted on a surface plate, horizontal and

Fig. 6-13
Checking the accuracy of the centre layout with a divider.

vertical lines can be laid out in one setup. Horizontal lines can be scribed when the angle plate is set on its base, while the vertical lines may be scribed when the angle plate is set on one end. A surface gauge may be set to a size or dimension by using a combination square, which is generally accurate enough for most layout work.

To Set a Surface Gauge to a Dimension (Fig. 6-15)

1. Thoroughly clean the top of the surface plate.
2. Set a combination square on the surface plate.
3. Loosen the combination square lock nut and be sure that the end of the rule is down against the surface plate.
4. Tighten the square lock nut.
5. Set the surface gauge on the surface plate.
6. Loosen the scriber clamp nut and adjust the scriber so that it is approximately at the desired dimension.
7. Tighten the scriber clamp nut securely to lock the scriber in position.
8. Turn the surface gauge thumb screw until the scriber point is in the centre of the desired graduation on the rule.

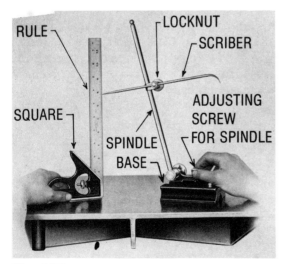

Fig. 6-15
Setting a surface gauge to a dimension using a combination square.

To Lay Out Horizontal Lines Using a Surface Gauge (Fig. 6-16)

1. Set the surface gauge to the dimension required.
2. Place the edge of the work from which the line is to be scribed on the surface plate.

Thin workpieces and those pieces requiring intersecting lines should be clamped to an angle plate (Fig. 6-16).

Courtesy Kostel Enterprises Ltd.

Fig. 6-16
Using the surface gauge to lay out lines parallel to the top of the surface plate.

3. Hold the surface gauge *down* on the surface plate.
4. Draw the surface gauge along the work in the direction of the arrow to scribe the line.

Pushing will cause the scriber point to dig into the work and cause an inaccurate layout.

5. Reset the surface gauge to the combination square rule for each dimension.
6. Lay out all lines that are parallel to the edge resting on the surface plate.
7. If intersecting lines are required on the layout, reset the left-hand clamp as shown in Fig. 6-17.
8. Place the angle plate on its left-hand end, set the surface gauge to the proper height, and scribe the intersecting lines (Fig. 6-18).

SURFACE GAUGE 43

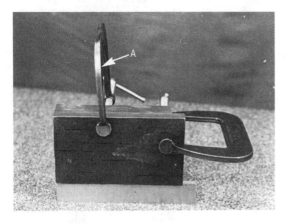

Fig. 6-17
When the work must be rotated 90°, the clamp "A" must be reset.

Fig. 6-18
Scribing the intersecting lines on a workpiece.

ANGLE PLATES

An angle plate (Fig. 6-19) is a precision L-shaped tool made of cast iron or hardened steel, machined to an accurate 90° angle, with all working surfaces and edges ground square and parallel. Angle plates are used to hold work parallel and at right angles to a surface. C-clamps are generally used to fasten work to an angle plate; however, some angle plates are provided with slots and tapped holes for this purpose.

Courtesy Kostel Enterprises Ltd.

Fig. 6-19
Work may be clamped to an angle plate for layout purposes.

On work where a number of horizontal and vertical lines are to be scribed, clamp the pieces to an angle plate and lay out all the horizontal lines. Without removing the clamps, set the angle plate on an edge and proceed to lay out the vertical lines. The horizontal and vertical lines will be at right angles to each other because all the edges on an angle plate are at right angles.

CLAMPS

C-clamps (Fig. 6-20A) are clamps made in the shape of a "C". They are used to fasten work to an angle plate or a drill press table and also for holding two or more pieces of metal together.

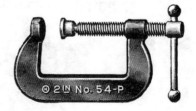

Courtesy Cincinnati Tool Co.

Fig. 6-20A
C-Clamp

44   CHAPTER 6 / LAYOUT TOOLS

Parallel clamps (Fig. 6-20B) are clamps consisting of two jaws held together with adjusting screws. They are used for the same purpose as C-clamps.

Courtesy Brown and Sharpe Manufacturing Co.

Fig. 6-20B
Parallel Clamp

PARALLELS

Parallels (Fig. 6-21) are square or rectangular hardened steel bars whose surfaces have been ground square and parallel. They are made in pairs and are used in layout work to raise the work to a suitable height and provide a solid seat. When used under work on a surface plate, the bottom work surface is held parallel to the top of the surface plate (Fig. 6-19).

Courtesy Taft-Peirce Manufacturing Co.

Fig. 6-21
Parallels are available in a variety of shapes and sizes.

V-BLOCKS

V-blocks (Fig. 6-22) are generally made of hardened steel or cast iron and are available in a wide range of sizes. They are used when laying out or drilling round work. Usually they come in pairs and have an accurate 90° V-shaped slot machined in the top and bottom. A U-shaped clamp is generally supplied with V-blocks in order to hold the work securely.

Fig. 6-22
V-blocks may be used when laying out round work.

RADIUS GAUGES

Radius gauges are widely used by machinists for checking and laying out concave and convex radii of all types. Radius gauges are manufactured individually, such as those illustrated in Fig. 6-23, or in a series of radius leaves mounted in a holder. Radius gauges are available in a wide range of sizes. Two standard metric sets are available. One set ranges from 0.75 to 5 mm radius in steps of 0.25 mm. Another set ranges from 5.5 to 13 mm in steps of 0.5 mm. The most common inch set consists of gauges from 1/64 in. to 33/64 in., varying in steps of 1/64 in. from one size to the next.

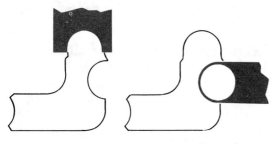

Courtesy Lufkin Rule Co. of Canada, Ltd.

Fig. 6-23
Radius gauges may be used to lay out or check convex or concave radii.

TEMPLATES

A template is usually made of a thin piece of metal, cut to the exact shape of the finished work. A template is used to lay out each piece when many workpieces of the same shape are required. The surface of the work is first coated with layout dye. The outline of the template is then traced on each workpiece using a sharp scriber. In this way all workpieces are laid out exactly the same. Templates may also be used to check the accuracy of contours or of special forms.

TEST YOUR KNOWLEDGE

1. Define the term "laying out."
2. Explain why all layout should be made from a "base" line or machined edge.

Preparing Surfaces for Layout

3. Why is layout material applied to the surface of a workpiece?
4. Why is it important that the work surface be clean before layout material is applied?
5. List three methods of coating work surfaces for layout work.

Surface Plates

6. Why are surface plates used in layout work?
7. How can the accuracy of surface plates be maintained?

Scriber

8. Of what type of material are scribers made?
9. Why must the point of a scriber *always* be sharp?
10. How should a scriber be held in order to scribe an accurate line?

Punches

11. Describe a prick punch and state the purpose for which it is used.
12. Why are prick punch marks sometimes called *witness* marks?
13. Describe a centre punch and state the purpose for which it is used.
14. List four points which should be kept in mind when using a prick or centre punch.

Divider

15. For what purpose are dividers used?
16. Explain how a divider should be set to a size.

Solid Square

17. Describe a solid square.
18. List three uses of a solid square.

Combination Set

19. Name the four main parts of a combination set.
20. How are the combination set rules usually graduated?
21. List three uses of a combination square.
22. How far should the steel rule extend beyond the body of the square when laying out parallel lines?
23. For what purpose are bevel protractors used?
24. For what purpose is the centre head used?

Hermaphrodite Calipers

25. For what purpose are hermaphrodite calipers used?

Centring Round Stock

26. List the steps necessary to lay out the centre of round work with the centre head.

27. List the steps required to lay out the centre of a round piece of stock using a hermaphrodite caliper.
28. What must be done to the ends of the workpiece before centring it with a bell-centre punch? Why?
29. Briefly explain how the accuracy of the centre layout may be tested with dividers.

Surface Gauge
30. Describe a surface gauge.
31. List three uses of a surface gauge.
32. List the steps involved in setting a surface gauge to a dimension using a combination square.
33. What procedure should be followed when scribing a line with a surface gauge?
34. How can an angle plate be used to lay out horizontal and vertical lines in one setup?

Angle Plates, Parallels, V-Blocks
35. Describe an angle plate.
36. Explain why an angle plate is a valuable tool for layout work.
37. What are parallels?
38. How are parallels used in layout work?
39. Describe a V-block.
40. List two uses for V-blocks.

CHAPTER 7
METALLURGY

Courtesy Kaiser Steel Corp.

IRON AND STEEL

In ancient times iron was a rare and precious metal. Today steel, a purified form of iron ore, has become one of humanity's most useful servants. Nature supplied the basic raw products of iron ore, coal, and limestone, and our ingenuity has converted them into a countless number of products. Steel can be made hard enough to cut glass, pliable as the steel in a paper clip, flexible as the steel in springs, or strong enough to withstand a stress measuring 3445 MPa (500,000 psi). It can be drawn into wire 0.02 mm (0.001 in.) thick or fabricated into giant girders for buildings and bridges. Steel can also be made resistant to heat, cold, rust, and chemical action. Steel is truly our most versatile metal.

RAW MATERIALS

The raw materials of steel making, *iron ore*, *coal*, and *limestone*, must be brought together, often from great distances, and smelted in a blast furnace to produce the pig iron that is used to make steel.

IRON ORE

Iron ore is the chief raw material used in the manufacture of iron and steel. In Canada, large ore deposits are found in the Steep Rock and Michipicoten districts on the north shore of Lake Superior and in the Ungava district near the Quebec-Labrador border. The main sources of iron ore in the United States are the Great Lakes states of Michigan, Minnesota, and Wisconsin.

Mining Iron Ore

When layers of iron ore are near the earth's surface, the surface material, consisting of sand, gravel, and boulders, is first removed. Then the iron ore is scooped up by power shovels and loaded into trucks or railway cars (Fig. 7-1). This is called *open pit mining*. About 75% of the ore mined is removed by this method.

When layers of iron ore are too deep in the earth to make open pit mining economical, *underground* or *shaft mining* is used. Shafts are sunk into the earth, and passageways are cut into the ore body. The ore is blasted loose and brought to the surface by shuttle cars or conveyor belts.

Types of Iron Ore

Some of the most important types of iron ore are:
HEMATITE, a rich ore containing about 70% iron. It ranges in colour from a grey to a bright red.
LIMONITE, a high-grade brown ore containing water, which must be removed before the ore can be shipped to steel mills.
MAGNETITE, a rich grey to black magnetic ore containing over 70% iron.

Courtesy Steel Company of Canada, Ltd.

Fig. 7-1
Open pit mining of iron ore.

TACONITE, a low grade ore containing only about 20 to 30% iron, which is uneconomical to use without further treatment.

Pelletizing Process

Low grade iron ores are uneconomical to use in the blast furnace and as a result go through a pelletizing process, where most of the rock is removed and the ore is brought to a higher iron concentration. Some steel-making firms are now pelletizing most of their ores to reduce transportation costs and the problems of pollution and slag disposal at the steel mills.

The crude ore is crushed and ground into a powder and passed through magnetic separators, where the iron content is increased to about 65% (Fig. 7-2A). This high grade material is mixed with clay and formed into pellets about 12 to 20 mm (1/2 to 3/4 in.) in diameter in a pelletizer. The pellets are then covered with coal dust and sintered (baked) at 1290°C. (2354°F) (Fig. 7-2B). The resultant hard, highly concentrated pellets will remain intact during transportation and loading into the blast furnace.

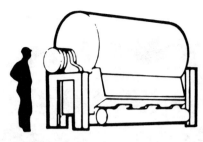

Fig. 7-2A
Iron ore is separated from the rock in a magnetic separator.

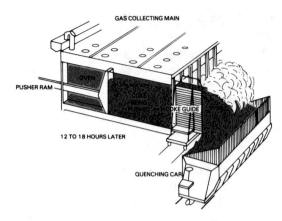

Courtesy American Iron and Steel Institute

Fig. 7-3
Coal is converted into coke in long, narrow coking ovens.

Fig. 7-2B
Iron ore pellets are sintered (hardened) in a pellet hardening furnace.

COAL

Coke, which is used in the blast furnace, is made from a special grade of soft coal that contains small amounts of phosphorus and sulphur. The main sources of this coal are mines in West Virginia, Pennsylvania, Kentucky, and Alabama. Coal may be mined by either strip or underground mining. Before being converted into coke, the coal is crushed and washed. It is then loaded into the top of long, high, narrow ovens which are tightly sealed to exclude air. The coal is baked at 1150°C (2150°F) for about eighteen hours, then dumped into rail cars and quenched with water (Fig. 7-3).

The gases formed during the coking process are distilled into valuable by-products such as tar, ammonia, and light oils. These by-products are used in the manufacture of fertilizers, nylon, synthetic rubber, dyes, plastic, explosives, aspirin, and sulpha drugs.

LIMESTONE

Limestone, a grey rock consisting mainly of calcium carbonate, is used in the blast furnace as a flux to fuse and remove the impurities from the iron ore. It is also used as a purifier in the steelmaking furnaces. Limestone is usually found fairly close to steelmaking centres. It is generally mined by open pit quarrying, where the rock is removed by blasting. It is crushed to size before shipment to the steel mill.

MANUFACTURE OF PIG IRON

The first step in the manufacture of any iron or steel is the production of *pig iron* in the blast furnace. The blast furnace, Fig. 7-4, about 40 m (130 ft.) high, is a huge steel shell lined with heat-resistant brick. Once started, the blast furnace runs continuously until the brick lining needs renewal or the demand for the pig iron drops.

Iron ore, coke, and limestone are measured out carefully and transported to the top of the furnace in a *skip car*, Fig. 7-4. Each ingredient is dumped separately into

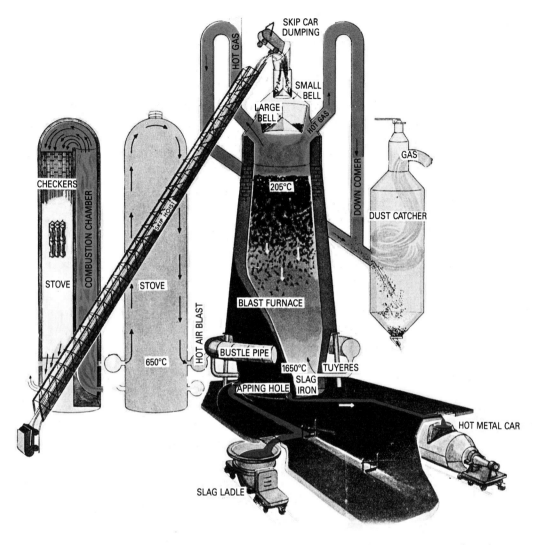

Fig. 7-4
A schematic view of a blast furnace.

Courtesy American Iron and Steel Institute

the furnace through the bell system, forming layers of coke, limestone, and iron ore in the top of the furnace. A continuous blast of hot air from the stoves at 650°C (1200°F) passes through the *bustle pipe* and *tuyeres* causing the coke to burn vigorously. The temperature at the bottom of the furnace reaches about 1650°C (3000°F) or higher. The carbon of the coke unites with the oxygen of the air to form carbon monoxide, which removes the oxygen from the iron ore and liberates the metallic iron. The molten iron trickles through the charge and collects in the bottom of the furnace.

The intense heat also melts the limestone, which combines with the impurities from the iron ore and coke to form

MANUFACTURE OF PIG IRON   51

a scum called slag. The slag also seeps down to the bottom of the charge and floats on top of the molten pig iron.

Every four to five hours the furnace is tapped and the molten iron, up to 315 t (350 tons), flows into a *hot metal* or *bottle car* and is taken to the steelmaking furnaces. Sometimes the pig iron is cast directly into *pigs*, which are used in foundries for making cast iron. At more frequent intervals, the slag is drawn off into a *slag car* or *ladle* and is later used for making mineral wool insulation, building blocks, and other products.

MANUFACTURE OF CAST IRON

Most of the pig iron manufactured in a blast furnace is used to make steel. However, some is also used to manufacture cast iron products. Cast iron is manufactured in a cupola furnace, which resembles a huge stove pipe (Fig. 7-5).

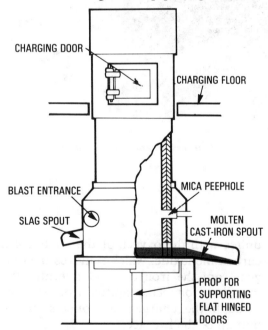

Fig. 7-5
A cupola furnace is used to manufacture cast iron.

Layers of coke, solid pig iron, scrap iron, and limestone are charged into the top of the furnace. After the furnace is charged, the fuel is ignited and air is forced in near the bottom to aid combustion. When the iron is melted, it settles to the bottom of the of the furnace and is then tapped into ladles.

The molten iron is poured into sand moulds of the required shape, and the metal assumes the shape of the mould. After the metal has cooled, the castings are removed from the moulds.

The principal types of iron castings are:

GRAY IRON CASTINGS, made from a mixture of pig iron and steel scrap, are the most widely used. They are made into a wide variety of products, including bathtubs, sinks, and parts for automobiles, locomotives, and machinery.

CHILLED IRON CASTINGS are made by pouring molten metal into metal moulds so that the surface cools very rapidly. The surface of such castings becomes very hard. They are used for crusher rolls and for other products requiring a hard, wear-resistant surface.

ALLOYED CASTINGS contain certain amounts of alloying elements such as chromium, molybdenum, and nickel. Castings of this type are used extensively by the automobile industry.

MALLEABLE CASTINGS are made from a special grade of pig iron and foundry scrap. After these castings have solidified, they are annealed in special furnaces. This makes the iron malleable and resistant to shock.

MANUFACTURE OF STEEL

Before molten pig iron from the blast furnace can be converted into steel, some of its impurities must be burned out. This is done in one of three different types of furnace: *the open hearth furnace, the basic oxygen furnace,* or *the electric furnace.*

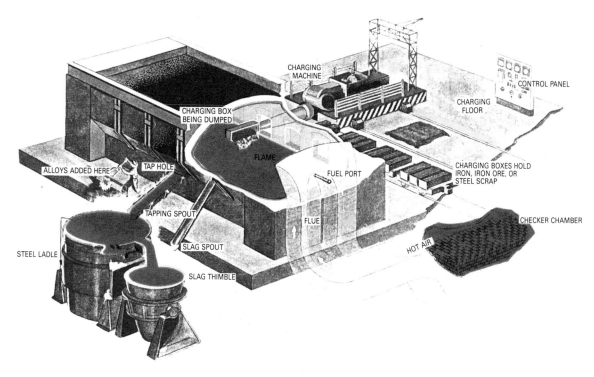

Courtesy American Iron and Steel Institute

Fig. 7-6
A schematic view of an open hearth furnace.

For many years, about 90% of all steel produced in North America was made by *open hearth furnaces*. The remainder of the steel was produced by *bessemer converters* and *electric furnaces*.

With the introduction of the *basic oxygen process* in 1955, the emphasis on steelmaking processes shifted. Steelmakers found that the addition of oxygen to any steelmaking process speeded production. The basic oxygen furnace was developed as a result. Today almost all steel produced is made in basic oxygen furnaces, with much of the remainder being produced in open hearth furnaces modified by the addition of oxygen lances. Special tool steels are still made by electric furnaces, but the production of steel by the Bessemer process has become practically non-existent.

OPEN HEARTH FURNACE

Open hearth furnaces (Fig. 7-6) are rectangular brick structures resembling huge ovens 12 to 15 m (40 to 50 ft.) long and 4.5 to 6.0 m (15 to 17 ft.) wide. Until 1960, about 90% of all steel manufactured was produced in open hearth furnaces. These furnaces are rapidly being replaced by the more efficient basic oxygen furnaces.

When a furnace is charged, limestone is put in first as a flux; then steel scrap is added. After the scrap has been melted down somewhat, molten pig iron is poured in. Tongues of flame sweep back and forth over the materials, creating temperatures of about 1650°C (3000°F), melting and burning out impurities.

Most open hearth furnaces have now been updated with the addition of the oxy-

MANUFACTURE OF STEEL 53

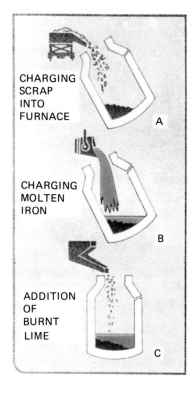

Fig. 7-7
The operation of the basic oxygen furnace.

Courtesy Inland Steel Corporation

THE BASIC OXYGEN FURNACE

The basic oxygen furnace (Fig. 7-7) is a cylindrical, brick-lined furnace with a dished bottom and cone-shaped top. It may be tilted in both directions for charging and tapping, but is kept in the vertical position during the steel-making process.

With the furnace tilted forward, scrap steel (30 to 40% of the total charge) is loaded into the furnace (Fig. 7-7A). Molten pig iron (60 to 70% of the total charge) is then added (Fig. 7-7B). The furnace is then moved to the vertical position where the fluxes (mainly burnt lime) are added (Fig. 7-7C). When the furnace is still in the upright position, a water-cooled oxygen lance is lowered into the furnace until the top of the lance is the required height above the molten metal. High pressure gen lance. This directs almost pure oxygen at a high velocity onto the top of the molten steel in the hearth. The oxygen burns the impurities out of the steel much quicker and can make a heat of 209 t (230 tons) of steel in approximately 6 hours, as compared to 7-1/2 hours without the oxygen lance.

Frequent samples are taken from the bubbling metal for chemical analysis. On the basis of the laboratory reports on these analyses, alloying materials are added to bring the molten metal to the required chemical composition.

The molten steel is poured into large cast iron forms called *ingot moulds*. Here the steel is allowed to solidify before being taken to the rolling mills to be formed into various shapes and sizes.

oxygen is blown into the furnace, causing a churning, turbulent action during which the undesirable elements are burned out of the steel (Fig. 7-7D). The oxygen blow lasts for about 20 minutes. The lance is then removed and the furnace is tilted to a horizontal position. The temperature of the metal is taken and samples of the metal are tested. If the temperature and the samples are correct, the furnace is tilted to the tapping position (Fig. 7-7E) and the metal is tapped into a ladle. Alloying elements are added at this time to give the steel its desired properties. An alloy is a combination of two or more metals designed to give desired properties. After tapping, the furnace is tilted in the opposite direction and to an almost vertical position to dump the slag into a slag pot (7-7F). About 272 t (300 tons) of steel can be produced in 1 hour in a large basic oxygen furnace.

THE ELECTRIC FURNACE

The electric furnace (Fig. 7-8) is used primarily to make fine alloy and tool steels. Because the heat, the amount of oxygen, and the atmospheric conditions can be accurately controlled in the electric furnace, it is used to make steels that cannot be readily produced in any other way.

Carefully selected steel scrap, containing smaller amounts of the alloying elements than are required in the finished steel, is loaded into the furnace. The three carbon electrodes are lowered until an arc strikes from them to the scrap. The heat generated by the electric arcs gradually melts all the steel scrap. Alloying materials, such as chromium, nickel, tungsten, etc., are then added to make the type of alloy steel required. Depending upon the size of the furnace, it takes from 4 to 12 hours to make a heat of steel. When the metal is ready to be tapped, the furnace is tilted forward and the steel flows into a large ladle. From the ladle, the steel is teemed into ingots.

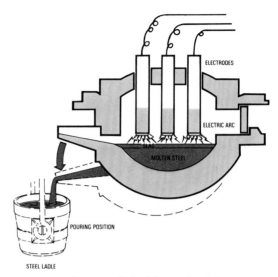

Courtesy United States Steel Corporation

Fig. 7-8
Electric furnaces are used to manufacture fine alloy and tool steels.

STEEL PROCESSING

After steel has been refined in any of the furnaces, it is tapped into ladles, where the necessary alloying elements and deoxidizers may be added. The molten steel may be teemed into ingots for later use or be formed directly into slabs by the *continuous-casting process*.

Steel teemed into ingot moulds is allowed to solidify. The ingot moulds are then removed or stripped, and the hot ingots are placed into soaking pits at 1204°C (2200°F) to bring them to a uniform temperature. The reheated ingots are then sent to rolling mills, where they are rolled into various shapes such as *blooms, billets, and slabs* (Fig. 7-9).

BLOOMS are generally rectangular or square and are larger than 232 cm² (36 sq. in.) in cross-section. They are used to manufacture structural steel and rails.

BILLETS may be rectangular or square, but are less than 232 cm² (36 sq. in.) in cross-sectional area. They are used to manufacture steel rods, bars, and pipes.

SLABS are usually thinner and wider than billets. They are used to manufacture plate, sheet, and strip steel.

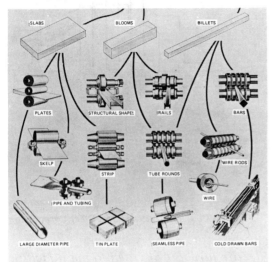

Fig. 7-9
Various shapes of steel are produced by rolling.

Strand or Continuous Casting

Strand or continuous casting, Fig. 7-10, is the most modern and efficient method of converting molten steel into semi-finished shapes such as slabs, blooms, and billets. Molten steel from the furnace is taken in a ladle to the top of the strand or continuous caster and poured into the tundish. The tundish acts as a reservoir, permitting the empty ladle to be removed and the full ladle to be poured without interrupting the flow of molten metal to the caster. The steel is stirred continuously by a nitrogen lance or by electromagnetic devices.

The molten steel drops in a controlled flow from the tundish into the mould section. Cooling water in the mould wall quickly chills the outside of the metal to form a solid skin, which becomes thicker as the steel strand descends through the cooling system. As the strand reaches the bottom of the machine, it becomes solid throughout. The solidified steel is moved in a gentle curve by bending rolls until it reaches a horizontal position. The strand is then cut by a travelling cutting torch into required lengths. In some strand casting machines, the solidified steel is cut when it is in the vertical position. The slab or billet then topples to the horizontal position and is taken away.

FERROUS METALS

The three general classes of ferrous metals are steel, cast iron, and wrought iron. Ferrous metals are made up principally of iron, which is magnetic. Steel is the most important ferrous metal used in machine shop work.

Types of Steel

LOW-CARBON STEEL, commonly called machine steel, contains from 0.10 to 0.30% of carbon. This steel, which is easily forged, welded, and machined, is used for making such things as chains, rivets, bolts, shafting, etc.

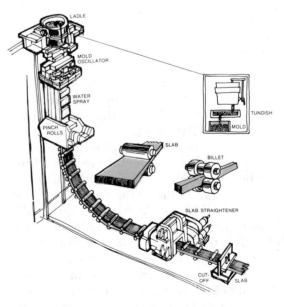

Courtesy American Iron and Steel Institute

Fig. 7-10
The continuous-casting process for manufacturing steel blooms or slabs.

MEDIUM-CARBON STEEL contains from 0.30 to 0.60% carbon and is used for heavy forgings, car axles, rails, etc.

HIGH-CARBON STEEL, commonly called tool steel, contains from 0.60 to 1.7% carbon and can be hardened and tempered. Hammers, crowbars, etc., are made from steel having 0.75% carbon. Cutting tools, such as drills, taps, reamers, etc., are made from steel having 0.90 to 1.00% carbon.

ALLOY STEELS are steels which have certain metals, such as chromium, nickel, tungsten, vanadium, etc., added to them to give the steel certain new characteristics. By the addition of various alloys, steel can be made resistant to rust, corrosion, heat, abrasion, shock, and fatigue.

HIGH-SPEED STEELS contain various amounts and combinations of tungsten, chromium, vanadium, cobalt, and molybdenum. Cutting tools made of such steels are used for machining hard materials at high speeds and for taking heavy cuts. High-speed steel cutting tools are noted for maintaining a cutting edge at temperatures where most steels would break down.

HIGH-STRENGTH, LOW-ALLOY STEELS contain a maximum carbon content of 0.28% and small amounts of vanadium, columbium, copper, and other alloying elements. They have higher strength than medium carbon steels and are less expensive than other alloy steels. These steels develop a protective coating when exposed to the atmosphere and therefore do not require painting.

Chemical Elements in Steel

CARBON in steel may vary from 0.01 to 1.7%. The amount of carbon will determine the brittleness, hardness, and strength of the steel.

MANGANESE in low-carbon steel makes the metal ductile and of good bending quality. In high-speed steel it toughens the metal and raises its critical temperature. Manganese content usually varies from 0.39 to 0.80%, but may run higher in special steels.

PHOSPHORUS is an undesirable element which makes steel brittle and reduces its ductility. In satisfactory steels the phosphorus content should not exceed 0.05%.

SILICON is added to steel in order to remove gases and oxides, thus preventing the steel from becoming porous and oxidizing. It makes the steel harder and tougher. Low-carbon steel contains about 0.20% silicon.

SULPHUR, an undesirable element, causes crystallization of steel (hot shortness) when the metal is heated to a red colour. Good quality steel should not contain more than 0.04% of sulphur.

PHYSICAL PROPERTIES OF METALS

To better understand the use of the various metals, one should be familiar with the following terms:

BRITTLENESS (Fig. 7-11A) is the property of a metal which permits no permanent distortion before breaking. Cast iron is a brittle metal; it will break rather than bend under shock or impact.

Courtesy Linde Division, Union Carbide Corporation

Fig. 7-11A
Brittle metals will not bend, but break easily.

DUCTILITY (Fig. 7-11B) is the ability of the metal to be permanently deformed without breaking. Metals such as copper and machine steel, which may be drawn into wire, are ductile materials.

Courtesy Linde Division, Union Carbide Corporation

Fig. 7-11B
Ductile metals can easily be deformed.

ELASTICITY (Fig. 7-11C) is the ability of a metal to return to its original shape after any force acting upon it has been removed. Properly heat-treated springs are good examples of elastic materials.

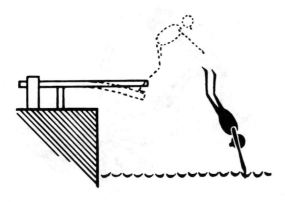

Courtesy Linde Division, Union Carbide Corporation

Fig.7-11C
Elastic metals return to their original shape after the load is removed.

HARDNESS (Fig. 7-11D) may be defined as the resistance to forceable penetration or plastic deformation.

Courtesy Linde Division, Union Carbide Corporation

Fig. 7-11D
Hard metals resist penetration.

MALLEABILITY (Fig. 7-11E) is that property of a metal which permits it to be hammered or rolled into other sizes and shapes.

Courtesy Linde Division, Union Carbide Corporation

Fig. 7-11E
Malleable metals may easily be formed or shaped.

TOUGHNESS is the property of a metal to withstand shock or impact. Toughness is the opposite condition to brittleness.

FATIGUE-FAILURE is the point at which a metal breaks, cracks, or fails as a result of repeated stress.

TABLE 7-1 IDENTIFICATION OF METALS

Metal	Carbon Content	Appearance	Method of Processing	Uses
Cast Iron (C.I.)	2.5 to 3.5%	Grey, rough sandy surface	Molten metal poured into sand moulds	Parts of machines, such as Lathe beds, etc.
Machine Steel (M.S.)	0.10 to 0.30%	Black, scaly surface	Put through rollers while hot	Bolts, Rivets, Nuts, Machine parts
Cold Rolled or Cold Drawn (C.R.S.) (C.D.S.)	0.10 to 0.30%	dull silver, smooth surface	Put through rollers or drawn through dies while cold	Shafting, Bolts, Screws, Nuts
Tool Steel (T.S.)	0.60 to 1.5%	Black, glossy	Same as Machine steel	Drills, Taps, Dies, Tools
High-Speed Steel (H.S.S.)	Alloy Steel	Black, glossy	Same as Machine steel	Dies, Tools, Taps, Drills, Toolbits
Brass	....	Yellow (various shades). Rough if cast, smooth if rolled	Same as Cast iron, or rolled to shape	Bushings, Pump parts, Ornamental work
Copper	....	Red-Brown. Rough if cast, smooth if rolled	Same as Cast iron, or rolled to shape	Soldering irons, Electric wire, Water pipes

Alloying Elements in Steel

Alloying elements may be added during the steelmaking process to produce certain qualities in the steel. Some of the more common alloying elements are:

CHROMIUM in steel imparts hardness and wear resistance. It gives steel a deeper hardness penetration than other alloying metals. It also increases resistance to corrosion.

MOLYBDENUM allows cutting tools to retain their hardness when hot. Because molybdenum improves steel's physical structure, it gives steel a greater ability to harden.

NICKEL in steel improves its toughness, resistance to fatigue-failure, impact properties, and resistance to corrosion.

TUNGSTEN increases the strength and toughness of steel and its ability to harden. It also gives cutting tools the ability to maintain a cutting edge even at a red heat.

VANADIUM in amounts up to 0.20% will increase steel's tensile strength and produce a finer grain structure in steel. Vanadium steel is usually alloyed with chromium to make such parts as springs, gears, wrenches, car axles, and many drop-forged parts.

IDENTIFICATION OF METALS

The machinist's work consists of machining metals, and therefore it is advantageous to learn as much as possible about the various metals used in the trade. It is often necessary to determine the type of metal being used by its physical appearance. Some of the more common machine shop metals and their appearance, use, etc., are found in Table 7-1. Metals are usually identified by one of four methods:
(a) by their appearance,
(b) by spark testing,
(c) by a manufacturer's stamp,
(d) by a code colour painted on the bar.

The latter two methods are most commonly used and are probably the most reliable. Each manufacturer, however, may use their own system of stamps or code colours.

SPARK TESTING

Any ferrous metal, when held in contact with a grinding wheel, will give off characteristic sparks. Small particles of metal, heated to red or yellow heat, are hurled into the air, where they come in contact with oxygen and oxidize or burn. An element such as carbon burns rapidly, resulting in a bursting of the particles. Spark bursts vary in colour, intensity, size, shape, and the distance they fly, according to the composition of the metal that is being ground.

LOW-CARBON OR MACHINE STEEL (Fig. 7-12A) shows sparks in long, light yellow streaks with a little tendency to burst.

MEDIUM-CARBON STEEL (Fig. 7-12B) is similar to low-carbon steel, but has more sparks which burst with a sparkler effect because of the greater percentage of carbon in the steel.

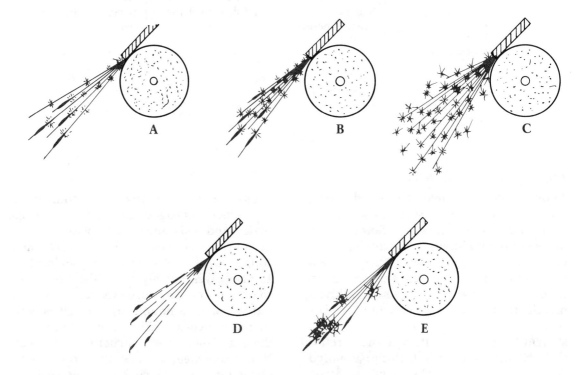

Fig. 7-12
Spark testing may be used to identify ferrous metals.

HIGH-CARBON OR TOOL STEEL (Fig. 7-12C) shows numerous little yellow stars bursting very close to the grinding wheel.
HIGH-SPEED STEEL (Fig. 7-12D) produces several interrupted spark lines with a dark red, ball-shaped spark at the end.
CAST IRON (Fig. 7-12E) shows a definite torpedo-shaped spark with a feather-like effect near the end. It changes from a dark red to a gold colour.

NONFERROUS METALS

Nonferrous metals are metals that contain little or no iron. They are resistant to corrosion and are non-magnetic. In machine shop work, nonferrous metals are used where ferrous metals would be unsuitable. The most commonly used nonferrous metals are aluminum, copper, lead, nickel, tin and zinc.
ALUMINUM is made from an ore called *bauxite*. It is a white, soft metal used where a light, non-corrosive metal is required. Aluminum is usually alloyed with other metals to increase its strength and stiffness. It is used extensively in aircraft manufacture because it is only one-third as heavy as steel.
COPPER is a soft, ductile, malleable metal which is very tough and strong. It is reddish in colour and second only to silver as an electrical conductor. Copper forms the basis of brasses and bronzes.
LEAD is a soft, malleable, heavy metal which has a melting point of about 327°C (620°F). It is corrosion-resistant and used for lining vats, tanks and for covering cables. It is also used for making such alloys as babbitt and solder.
NICKEL is a very hard, corrosion-resistant metal. It is used as a plating agent on steel and brass, and is added to steel to increase its strength and toughness.
TIN is a soft, white metal having a melting point of about 232°C (450°F). It is very malleable and corrosion resistant. It is used in the manufacture of tin plate and tin foil. It is also used for making such alloys as babbitt, bronze, and solder.
ZINC is a bluish-white element which is fairly hard and brittle. It has a melting point of about 420°C (790°F) and is used mainly to galvanize iron and steel.

Nonferrous Alloys

A nonferrous alloy is a combination of two or more nonferrous metals completely dissolved in each other. Nonferrous alloys are made when certain qualities of both original metals are desired. Some common nonferrous alloys are:
BRASS is an alloy of approximately 2/3 copper and 1/3 zinc. Sometimes 3% lead is added to make it easy to machine. Its colour is normally a bright yellow, but this varies slightly according to the amounts of alloys it contains. Brass is widely used for small bushings, plumbing and radiator parts, fittings for water cooling systems, and miscellaneous castings.
BRONZE is an alloy composed mainly of copper, tin, and zinc. Some types of bronze contain such additions as lead, phosphorus, manganese, and aluminum to give them special qualities. Bronze is harder than brass and resists surface wear. It is used for machine bearings, gears, propellers, and miscellaneous castings.
BABBITT is a soft, greyish-white alloy of tin and copper. Antimony may be added to make it harder, and lead is usually added if a softer alloy is required. Babbitt is used in bearings of many reciprocating engines.

SHAPES AND SIZES OF METALS

Due to the wide variety of work performed in a machine shop and the necessity of conserving machining time as well as reducing the amount of metal cut into steel chips, metals are manufactured in a wide variety of shapes and sizes. When ordering steel for work that must be machined, it is recommended that it be purchased a little

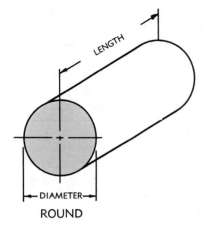
ROUND

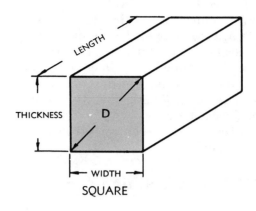
SQUARE

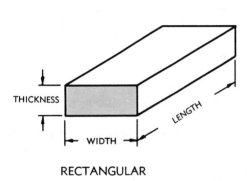
RECTANGULAR

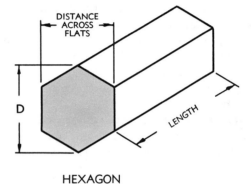
HEXAGON

Fig. 7-13
Shapes and sizes of metals.

larger than the finished size to allow for the machining operation. Although many factors determine exactly how much larger the piece should be, generally 1.5 mm (1/16 in.) oversize on each surface to be machined will prove adequate. For example, if a piece of round work must be finished to 19 mm (3/4 in.) diameter, it is wise to purchase 22 mm (7/8 in.) diameter material.

There is a proper method for specifying the sizes and dimensions of metals when ordering (see Fig. 7-13).
1. *Round material* has only two dimensions. Therefore when ordering, specify the diameter first and then the length.
2. *Flat or rectangular material* has three dimensions: thickness, width, and length, and should be ordered in that sequence.
3. *Square material* has three dimensions; however, the thickness and width are the same. When ordering, specify the thickness (or width) and then the length.
4. *Hexagonal material* has only two dimensions, the distance across flats and the length, and should be ordered in that sequence.

TEST YOUR KNOWLEDGE

1. Why is steel called our most versatile metal?

Raw Materials

2. Where are the chief sources of iron ore found in North America?

3. Name and briefly describe two methods of mining iron ore.
4. Name and describe three types of iron ore.
5. Briefly describe the pelletizing process for iron ore.
6. How is coal converted into coke?
7. What purpose does limestone serve in the steelmaking process?

Manufacture of Pig Iron

8. What type of furnace is used to manufacture pig iron?
9. List the raw materials used to manufacture pig iron.
10. Briefly describe the operation of a blast furnace.

Manufacture of Cast Iron

11. In what type of furnace is cast iron manufactured?
12. Explain the operation of a cupola furnace.
13. Name four types of cast iron and give one use for each.

Manufacture of Steel

14. Name three furnaces that are used to convert pig iron into steel.
15. Describe the operation of an open hearth furnace.
16. What effect did the development of the basic oxygen process have on steelmaking?
17. What are ingot moulds and what purpose do they serve?
18. Describe the basic oxygen furnace.
19. List the main steps in the operation of the basic oxygen furnace.
20. Explain why the electric furnace is used to produce fine alloy and tool steels.
21. List the main steps involved in producing steel in an electric furnace.
22. Briefly describe the continuous casting process used to convert molten steel into slabs or billets.

Ferrous Metals

23. Define a ferrous metal.
24. Give the carbon content and one use for each of the following:
    (a) low-carbon steel
    (b) medium-carbon steel
    (c) high-carbon steel
25. What are the advantages of alloy steels?
26. Describe the composition of high-speed steels.
27. Why is high-speed steel especially valuable for the manufacture of cutting tools?
28. Explain how carbon affects steel.
29. What two alloys help cutting tools maintain their hardness and cutting edge when hot?

Identification of Metals

30. Name four methods of identifying metals.
31. Describe the appearance of the following metals and give one use for each.
    (a) cast iron
    (b) machine steel
    (c) tool steel
    (d) brass
32. Why do sparks from different materials vary?
33. What are the spark characteristics of:
    (a) low-carbon steel?
    (b) high-carbon steel?
    (c) high-speed steel?

Nonferrous Metals

34. Define nonferrous metals and state why they are used in machine shop work.
35. Briefly describe and give one use for:
    (a) aluminum
    (b) copper
    (c) nickel
    (d) tin
36. What is a nonferrous alloy?
37. Describe the composition of: brass, bronze, babbitt.

Shapes and Sizes of Metals

38. Why are metals manufactured in a wide variety of shapes and sizes?
39. How much larger than the finished piece should work that requires machining be ordered?
40. Explain how round and flat material should be ordered.

CHAPTER 8
HAND TOOLS

At one time it was very important for a machinist to be highly skilled in the use of hand tools. The master tradespeople of the early twentieth century were known for the high skill and craftsmanship they had developed with hand tools. As newer and more accurate machine tools were developed, there was less need for the hand operations of old. Today it is important that we recognize the fact that if an operation can be performed on a machine, it will be done faster and more accurately.

However, hand tools are still essential for some machine shop operations, such as sawing, filing, polishing, tapping, and threading. It is still important that the apprentice, through patience and practice, become skilled in the use of hand tools. Hand tools must be used with the same care as is given to the more expensive machine tools. A reasonable amount of care will keep tools in safe and good working condition. One sign of a good machinist is the excellent condition of his or her tools.

THE MACHINIST'S VISE

The machinist's vise is a work-holding device used to hold work for such operations as sawing, filing, chipping, tapping, threading, etc.

Vises are made in a great variety of sizes, to hold work of many sizes and shapes. Some vises are equipped with a swivel base, which allows the vise to be turned to any position (Fig. 8-1). To hold finished work, it is wise to put jaws made of aluminum, brass, or copper over the regular jaws to protect the work.

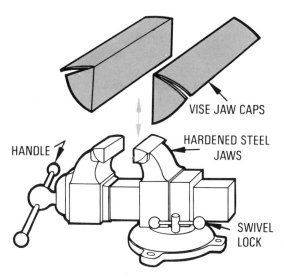

Fig. 8-1
A swivel base permits the work to be turned to any position in a 360° arc.

HAMMERS

The *ball-peen hammer*, or machinist's hammer as it is more commonly called, is the hammer generally used in machine shop work. The rounded top (Fig. 8-2) is called the peen, and the bottom is known as the face. Ball-peen hammers are made in a variety of sizes, with head masses ranging from approximately 110 to 790 g (4 to 28 oz.). They are hardened and tempered. The smaller sizes are used for layout work, while the larger ones are used for general bench work.

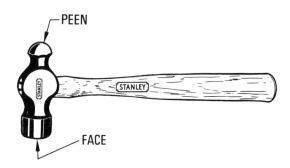

Courtesy Stanley Tools Division, Stanley Works

Fig. 8-2
A ball-peen hammer is the most common hammer used in machine shop.

Soft-faced hammers (Fig. 8-3) are used in assembly and setup work because they will not mar the finished surface of work. These hammers have pounding surfaces made of brass, plastic, lead, rawhide, or hand rubber.

When using a hammer, it should be grasped at the end of the handle. This position provides greater striking force and balance for the hammer than if it is

Courtesy Stanley Tools Division, Stanley Works

Fig. 8-3
A soft-faced hammer will not mar the finished surface of a workpiece.

gripped near the head. It also helps to keep the hammer face flat on the work being struck while minimizing the chance of damage to the face of the work.

The following safety precautions should be observed when using a hammer:
1. *A hammer with a loose head is dangerous.* Always keep the hammer handle tightly secured in the head with a suitable wedge.
2. *Replace the handle on any hammer if it is cracked* or does not look sound; do not wait for a serious accident to happen.
3. *Never use a hammer with a greasy handle or when hands are oily.*

METAL STAMPS

Metal stamps or stencils (Fig. 8-4) are used to mark or identify work pieces. They are made in a variety of sizes with letters or numbers from 0.8 to 12.7 mm (1/32 to 1/2 in.) high. Metal stamps should *never* be used on hardened metal, and if used on cast iron or hot rolled steel, the hard outer scale of the metal should first be removed by grinding, chipping, or machining, to prevent damage to the stamps.

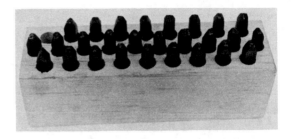

Courtesy Kostel Enterprises Ltd.

Fig. 8-4
Metal stamps may be used to identify workpieces.

To Use Metal Stamps
1. Place the work in a vise or on a flat bench plate.
2. Lay out a base line to indicate where the stamping is to be located.
3. Lay out a line for the centre of the lettering.
4. Hold the metal stamp so that the letter or trade mark on its side is facing you. This will ensure that the letter will be imprinted "right-side up".
5. Place the edge of the metal stamp on the layout line at the centre of the layout.
6. Stamp the middle letter on the centre line.

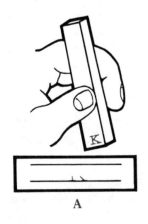 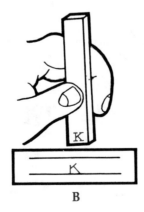 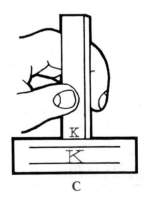

A   B   C

Fig. 8-5
The three-stamp method is used for impressing large metal stamps.

7. Stamp all the letters to the right of the centre line and then work from the centre to the left, thus balancing the stamping about the centre line.

When using metal stamps larger than 6.35 mm (1/4 in.), the three-step method is advisable in order to produce clear, sharp impressions.

To Use the 3-Step Method for Large Metal Stamps

1. Place the stamp on the base line inclined toward you and strike it sharply with a medium-sized ball-peen hammer (Fig. 8-5A).
2. Replace the stamp in the impression, inclined only slightly, and strike it sharply (Fig. 8-5B).
3. Replace the stamp in the impression, hold it *vertically*, and strike it sharply to complete the letter (Fig. 8-5C).

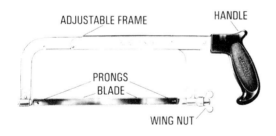

Courtesy L.S. Starrett Co.

Fig. 8-6
The main parts of a hand hacksaw.

HAND HACKSAW

The hacksaw is a hand tool used to cut metal. The pistol grip hacksaw (Fig. 8-6) consists of four main parts: handle, frame, blade, and adjusting wing nut. The frame on most hacksaws may be flat or tubular. Some hacksaws have adjustable frames to accommodate various hacksaw blade lengths.

HACKSAW BLADES

Hacksaw blades are made of high-speed, molybdenum, or tungsten alloy steel that has been hardened and tempered. The saw blades generally used are 12.7 mm (1/2 in.) wide, and 0.63 mm (1/4 in.) thick. Common lengths of hacksaw blades are 200, 250, and 300 mm, or 8, 10, and 12 in. There is a hole at each end of the blade for mounting it on the hacksaw frame.

The distance between each tooth on a blade is called the *pitch*. A pitch of 1/18 represents 18 teeth per inch. The most common blades have 14, 18, 24, or 32 teeth per inch. An *18-tooth blade* (18 teeth per inch) is recommended for general use. It is important to use the right pitch for the work being cut. Select a blade as coarse as possible in order to provide plenty of chip clearance and cut through the work quickly. The blade selected should have at least two teeth in contact with the work so that the work cannot jam between the teeth and strip the teeth from the saw blade.

Mounting a Blade on a Hand Hacksaw

1. Select the proper blade for the job (Fig. 8-7).
2. Adjust the frame to the length of the blade.
3. Place one end of the blade on the back pin (near the wing nut).
   NOTE: *Be sure that the teeth of the blade point away from the handle.*
4. Place the other end of the blade on the front pin.
5. Tighten the wing nut until the blade is just snugged up (Fig. 8-8).
   NOTE: *Do not tighten the blade too much, as the blade may be broken or the hacksaw frame bent.*

To Use a Hand Hacksaw

1. Check that the pitch is proper for the job and be sure that the teeth point away from the handle.
2. Adjust the tension on the saw blade as

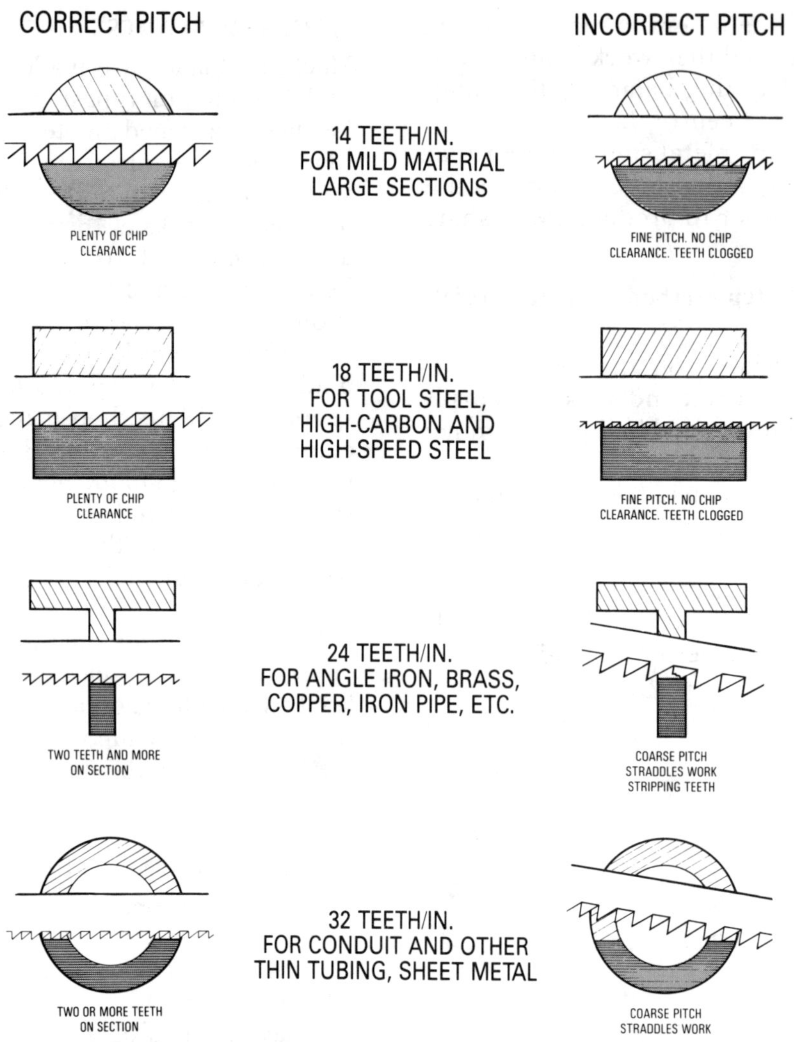

CORRECT PITCH   **INCORRECT PITCH**

14 TEETH/IN.
FOR MILD MATERIAL
LARGE SECTIONS

PLENTY OF CHIP CLEARANCE   FINE PITCH. NO CHIP CLEARANCE. TEETH CLOGGED

18 TEETH/IN.
FOR TOOL STEEL,
HIGH-CARBON AND
HIGH-SPEED STEEL

PLENTY OF CHIP CLEARANCE   FINE PITCH. NO CHIP CLEARANCE. TEETH CLOGGED

24 TEETH/IN.
FOR ANGLE IRON, BRASS,
COPPER, IRON PIPE, ETC.

TWO TEETH AND MORE ON SECTION   COARSE PITCH STRADDLES WORK STRIPPING TEETH

32 TEETH/IN.
FOR CONDUIT AND OTHER
THIN TUBING, SHEET METAL

TWO OR MORE TEETH ON SECTION   COARSE PITCH STRADDLES WORK

Fig. 8-7
Recommended saw pitches for various types of work.

tight as two-finger pressure will permit. *NOTE: Too much pressure will bend the frame and damage the blade. Too little pressure will allow the blade to bend, produce an inaccurate cut, and may break the blade.*

3. Mark the position of the cut on the workpiece with a layout line or a nick with a file.
4. Mount the stock in the vise so that the cut will be made about 6 mm (1/4 in.) from the vise jaws.
5. Grip the hacksaw as shown in Fig. 8-9A and assume a comfortable stance.
6. Position the blade on the work just outside the layout line or in the file nick.
7. Apply down-pressure on the forward stroke and release it on the return stroke. Use a speed of about 50 strokes/min.

If the cut does not start in the proper place, file a V-shaped nick at the cut-off mark to guide the saw blade.

8. When nearing the end of the cut, slow

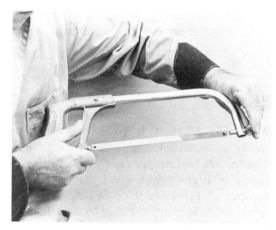

Courtesy Kostel Enterprises Ltd.

Fig. 8-8
When mounting a blade on a hand hacksaw, be sure that the teeth point away from the handle.

Courtesy Kostel Enterprises Ltd.

Fig. 8-9A
The correct method of holding a hand hacksaw.

Courtesy Kostel Enterprises Ltd.

Fig. 8-9B
Sawing thin metal between two pieces of wood.

down and ease up on the down-pressure to control the saw as it breaks through the material.

9. When cutting thin material, hold the saw at an angle so that at least two teeth will bear on the work at all times. Sheet metal and other thin material may be clamped between two thin pieces of wood. The cut is then made through all three pieces, as in Fig. 8-9B.

NOTE: Starting a new blade in an old cut will bind and ruin the "set" of the new blade. It is advisable to turn the work and start a cut in another place.

CHISELS

Modern machine tools have almost eliminated the use of chisels in machine shop work. However, there are times when the use of a chisel is invaluable.

The *flat cold chisel* (Fig. 8-10) is most commonly used for *chipping*. It is also used for cutting thin metals, cutting off the heads of rivets, and removing weld spots.

The *cape chisel* has a narrow cutting edge and is used for cutting keyways and narrow grooves in metal.

The *round nose or grooving chisel* has a rounded cutting edge and is used for cutting oil grooves in bearings and cutting grooves in curved and flat surfaces.

The *diamond point chisel* has a diamond-shaped cutting face. It is used for squaring corners and cutting V-shaped grooves.

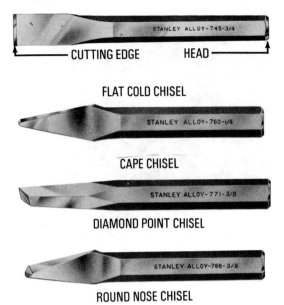

Courtesy Stanley Tools Division, Stanley Works

Fig. 8-10
Common types of cold chisels.

Hints on Chisel Use (Chipping)

1. Always wear safety goggles while chipping.
2. Never use a chisel with a mushroomed head.
3. Make sure the hammer and chisel head are free from grease and oil.
4. Place a block under the work in a vise to prevent it from slipping down during the chipping operation.
5. Hold the chisel firmly enough to guide it, but with the finger muscles relaxed.
6. Always chip toward the solid jaw of the vise, if possible.
7. Hold the hammer handle near the end and strike the chisel with quick sharp blows.
8. Never try to take too deep a cut. The angle at which the chisel is held determines the depth of cut.
9. When striking a chisel, watch the cutting edge and not the head of the chisel.
10. When chipping cast iron, always chip from the edge of the workpiece to the centre so that the edge will not break off.

FILES

A file (Fig. 8-11) is a hand-cutting tool with many teeth, used to remove surplus metal and produce finished surfaces. Files are made of high-carbon steel, hardened and tempered. Files are used in machine shop work when it is impractical to use machine tools. The file is a useful tool for removing machine marks, tool and die making, and the fitting of machine parts. Files are manufactured in a variety of shapes and sizes, each having a specific purpose. Files are divided into two classes, single- and double-cut files.

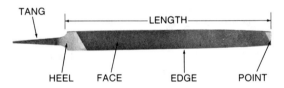

Fig. 8-11
The main parts of a hand file.

SINGLE-CUT FILES

These files have a single row of parallel teeth across the face at an angle from 65 to 85° (Fig. 8-12). Single-cut files are used when a smooth surface is desired, and when harder metals are to be finished.

DOUBLE-CUT FILES

These files have two rows of teeth crossing each other, one row being finer than the other (Fig. 8-12). The two rows crossing each other produce hundreds of sharp cutting teeth, which remove metal quickly and make for easy clearing of chips.

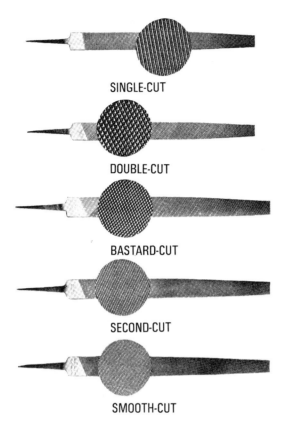

Fig. 8-12
Single- and double-cut files are manufactured in several degrees of coarseness.

DEGREES OF COARSENESS

Both single-cut and double-cut files are manufactured in various degrees of coarseness. On larger files this is indicated by the terms *rough, coarse, bastard, second-cut, smooth,* and *lead smooth.* The bastard, second-cut, and smooth files are the ones most commonly used in machine shops. On smaller files, the degree of coarseness is indicated by numbers from 00 to 8, number 00 being the coarsest.

SHAPES OF FILES

Files are manufactured in many shapes (Fig. 8-13) and may be identified by their cross-section, shape, or special use. The types of files most commonly used in a machine shop are the *mill, hand, flat, round, half-round, square, three-square (triangular), pillar, warding,* and *knife.*

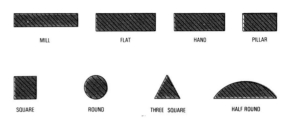

Fig. 8-13
Cross-sections of some common machinist's files.

CARE OF FILES

Proper care, selection, and use are important if good results are to be obtained when using a file. In order to preserve the life of a file, the following points should be observed:

1. Use a *file card* to keep the file clean and free of chips.
2. Do not knock a file on a vise or other metallic surface to clean it.
3. Do not apply too much pressure to a new file.
   This breaks down the cutting edges quickly. Too much pressure also causes "pinning" (small particles of metal become wedged between the teeth), which will cause scratches on the surface of the work.
4. Never use a file as a pry or hammer.
   NOTE: A file is a hardened tool which can snap easily. This may cause small pieces to fly and cause a serious eye injury.
5. Always store files where they will not rub together. Hang or store them separately.

Hints on Using Files

The following general points should be observed when filing.

1. *Never use a file without a handle.*

NOTE: *This practice can result in a serious injury. If the file should slip, the sharp tang could puncture a hand or arm.*

2. Always be sure that the handle is tight on the file.
3. To produce a flat surface when crossfiling, the right hand, forearm, and the left hand should be held in a horizontal plane (Fig. 8-14). Do not rock the file, but push it across the work in a straight line.

Courtesy Kostel Enterprises Ltd.

Fig. 8-14
The correct method of holding a file for general filing.

4. A file cuts only on the forward stroke. When filing, therefore, apply down-pressure on the forward stroke only and relieve the pressure on the return stroke.

   Applying down-pressure on the return stroke tends to dull a file.
5. Work to be filed should be held in a vise at about elbow height.

   Small, fine work may be held higher, while heavier work, requiring much removal of metal, should be held lower.
6. Never rub the hand or fingers across a surface being filed.

   Grease or oil from a hand deposited on the metal causes the file to slide instead of cutting the work. Oil will also cause the filings to clog the file. To prevent clogging, clean the file with a file card and then rub chalk over the surface of the file.

Filing Practice

At one time, filing was an important skill which every good machinist had to master. During the twentieth century the need for filing has decreased because new machine tools are able to perform the operations usually done by hand faster and more accurately, while at the same time producing very flat surfaces with a high surface finish. In the latter part of the twentieth century, it would be frowned upon if a person in the trade used a file for an operation which could be performed on a machine.

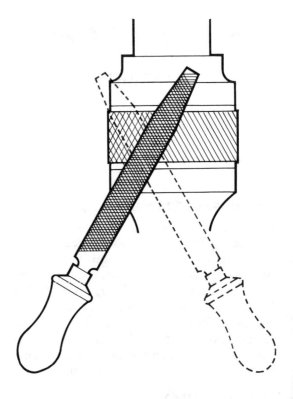

Fig. 8-15
Crossing the file stroke helps to maintain a flat surface.

Courtesy Kostel Enterprises Ltd.

Fig. 8-16
Use only short strokes and maintain down pressure on the file directly over the work when finish filing.

However, there are still times when it is necessary or more convenient to use a file, and it is wise for a machinist to be able to cross file and draw file.

Cross filing is used when metal is to be removed rapidly or the surface brought flat, prior to finishing by draw filing.

For rough filing, use a double-cut file and cross the stroke at regular intervals to help keep the surface flat and straight (Fig. 8-15). When finishing, use a single-cut file and take somewhat shorter strokes in order to keep the file flat on the work surface. The down-pressure on the file should be applied with the fingers of the left hand, which should be kept over the workpiece (Fig. 8-16) during the finishing operation.

The work should be tested occasionally for flatness by laying the edge of a steel rule across its surface (Fig. 8-17A). A steel square should be used to test the squareness of one surface to another (Fig. 8-17B).

Draw filing is used to produce a straight, square surface with a finer finish

Courtesy Kostel Enterprises Ltd.

Fig. 8-17B
Testing the workpiece for squareness with an adjustable square.

than is produced by straight filing. A single-cut file is used and pressure is applied to the file just above the edge of the workpiece (Fig. 8-18). The file is alternately pulled and pushed lengthwise along the

Courtesy Kostel Enterprises Ltd.

Fig. 8-17A
Testing the surface of a workpiece for flatness with the edge of a rule.

Courtesy Kostel Enterprises Ltd.

Fig. 8-18
Finger pressure should not extend beyond the edge of a workpiece when draw filing.

surface of the work. The file should be held flat and pressure applied on the forward stroke only. On the return stroke, the file should be slid back without applying pressure but without lifting the file from the work.

HAND TAPS

Taps are cutting tools used to cut internal threads. They are made of high-quality tool steel, high-speed steel, and various alloy steels. Special treatments, such as the titanium nitride coating of drills and taps, allow them to be run at higher speeds and feeds, thereby increasing productivity as much as tenfold. This coating also improves the life of the cutting tools, while at the same time producing superior finishes on holes and threads.

The most common taps have three or four flutes cut lengthwise across the threads to form cutting edges, provide clearance for the chips, and admit cutting fluid to lubricate the tap. The end of the shank is square so that a tap wrench can be used to turn the tap into a hole.

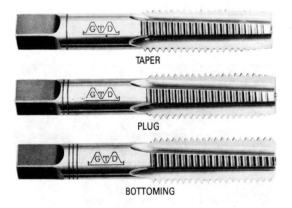

Courtesy Firth-Brown Tools (Canada) Ltd.

Fig. 8-19
A set of hand taps consists of a taper, plug, and bottoming tap.

Hand taps are usually made in sets of three, called *taper, plug, and bottoming* (Fig. 8-19).

A *taper tap* is tapered from the end approximately six threads and is used to start a thread easily. It can be used for tapping a hole which goes *through* the work, as well as for starting a *blind* hole (one that does not go through the work).

A *plug tap* is tapered for approximately three threads. Sometimes the plug tap is the only tap used to thread a hole going through a piece of work.

A *bottoming tap* is not tapered, but chamfered at the end for one thread. It is used for threading to the bottom of a blind hole. When tapping a blind hole, first use the taper tap, then the plug tap, and complete the hole with a bottoming tap.

INCH TAPS

Inch taps are available in a large variety of sizes, thread pitches, and thread forms. (See the Appendix, Table 5.) The major diameter, number of threads per inch, and type of thread are usually found stamped on the shank of a tap. For example: 1/2 in. — 13 N.C. represents:
(a) 1/2 in. = major diameter of the tap,
(b) 13 = number of threads per inch,
(c) N.C. = National Coarse (a type of thread).

METRIC TAPS

The International Standards Organization (ISO) has developed a standard metric thread which will be used in Canada, the United States, and many other countries throughout the world. This series will have only 25 thread sizes, ranging from 1.6 to 100 mm diameter. (See the Appendix, Table 6.)

Metric taps are identified with the letter **M** followed by the nominal diameter of the thread in millimetres times the pitch in millimetres. A tap with the markings M 2.5 × 0.45 would indicate:

M — a metric thread
2.5 — the nominal diameter of the thread in millimetres
0.45 — the pitch of the thread in millimetres.

TAP WRENCHES

Tap wrenches are manufactured in two types and in various sizes to suit the size of the tap being used.

The *double-end adjustable tap wrench* (Fig. 8-20A) is manufactured in several sizes, but is generally used for larger taps and in open places where there is room to turn the tap wrench. Because of the greater leverage obtained when using this wrench, it is important that a large wrench is not used to turn a small tap, as small taps are easily broken.

Courtesy Firth-Brown Tools (Canada) Ltd.

Fig. 8-20A
A double-end adjustable tap wrench.

The *adjustable T-handle tap wrench* (Fig. 8-20B) is generally used for small taps or in confined areas where it is not possi- tap to produce 75% of a full thread.

When used with very small number taps, the body of the wrench should be turned with the thumb and forefinger to advance the tap into the hole. The handle is usually used when threading with larger taps, which will not break as easily as small ones. *Always use the proper size tap wrench for the size of tap being used.*

TAP DRILL SIZE

Before a tap is used, the hole must be drilled in the workpiece to the correct tap drill size, Fig. 8-21. The tap drill size is the size of the drill that should be used to leave the proper amount of material in the hole for a tap to cut a thread. The tap drill, which is always smaller than the tap, leaves enough material in the hole for the tap to produce 75% of a full thread.

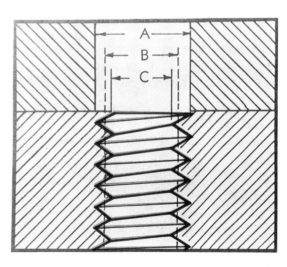

Courtesy Kostel Enterprises Ltd.

Fig. 8-21
Cross-section of a tapped hole.
A = body size
B = tap drill size
C = minor diameter

Courtesy Firth-Brown Tools (Canada) Ltd.

Fig. 8-20B
A T-handle tap wrench.

HAND TAPS

Tap Drill Sizes for Inch Threads

When a chart is not available, the tap drill size for any American National or Unified thread can be easily found by applying this simple formula:

$$T.D.S. = D - \frac{1}{N}$$

T.D.S. = tap drill size
D = major diameter of tap
N = number of threads per inch

EXAMPLE: Find the tap drill size for a 5/8 in.-11 UNC tap.

$$T.D.S. = \frac{5}{8} - \frac{1}{11}$$
$$= 0.625 - 0.091$$
$$= 0.534 \text{ in.}$$

The nearest drill size to 0.534 in. is 0.531 in. (17/32 in.). Therefore 17/32 in. is the tap drill size for a 5/8 in.-UNC tap. See the Appendix, Table 5, for a complete tapping drill chart for American National and Unified threads.

Tap Drill Sizes for Metric Threads

The tap drill sizes for metric threads may be calculated by subtracting the pitch from the nominal diameter.

$$T.D.S. = D - P$$

EXAMPLE: Calculate the tap drill size for a M12 × 1.75 thread.

$$T.D.S. = 12 - 1.75$$
$$= 10.25 \text{ mm}$$

Refer to the Appendix, Table 6, for tap drill sizes for metric threads.

TAPPING A HOLE

Tapping is the operation of cutting an internal thread using a tap and tap wrench. Because taps are hard and brittle, they are easily broken. *Extreme* care must be used when tapping a hole to prevent breakage. A broken tap in a hole is very difficult to remove and often results in scrapping the work.

Some of the most common causes of tap breakage and how to correct them are as follows:
1. The tap drill hole is too small.
   Be sure to drill the correct-size tap drill hole.
2. The tap's cutting edges are dull.
   Check each tap before use to be sure the cutting edges are sharp.
3. The tap was started out of alignment.
   Check the tap for squareness after it has entered the hole two full turns, and correct the alignment.
4. Too much pressure applied on one side of the tap wrench while trying to align the tap.
   Remove the tap from the hole and apply only slight pressure on the wrench while aligning the tap. Repeat as often as necessary until the tap is square.
5. Chips clog in the flutes while tapping.
   Clean the hole and tap of chips occasionally, and use cutting fluid while tapping.

To Tap a Hole by Hand

1. Select the correct size and type of tap for the job.
   (a) Use a plug tap for through holes.
   (b) Use the taper, plug, and bottoming tap in that sequence when tapping blind holes.
2. Select the correct tap wrench for the size of tap being used.
   CAUTION: Too large a tap wrench can result in a broken tap.
3. Apply a suitable cutting fluid.
   No cutting fluid is required when tapping brass or cast iron.
4. Place the tap in the hole as near to vertical as possible.
5. Apply equal down-pressure on both handles, and turn the tap clockwise (for right-hand thread) for about two turns.
6. Remove the tap wrench and check the tap for squareness.

Courtesy Kostel Enterprises Ltd.

Fig. 8-22
Check the tap for squareness at two positions 90° apart.

Check at two positions 90° to each other (Fig. 8-22).
7. If the tap has not entered squarely, remove it from the hole and restart it by applying *slight* pressure in the direction from which the tap leans (Fig. 8-23).
Be careful not to exert too much pressure in the straightening process; otherwise the tap may be broken.
8. When a tap has been properly started, feed it into the hole by turning the tap wrench. Down-pressure is no longer required, since the tap will thread itself into the hole.
9. Turn the tap clockwise one-quarter of a turn and then turn it backward about one-half turn to break the chip. This must be done with a steady motion to avoid breaking the tap.

When tapping blind holes, use all three taps in order: taper, plug, and then the bottoming tap. Before using the bottoming tap, remove all the chips from the hole and be careful not to hit the bottom of the hole with the tap.

TAPPING LUBRICANTS

The use of a suitable cutting lubricant when tapping results in longer tool life, better finish, and greater production. The recommended tapping lubricants for the more common metals are listed below. A more complete list will be found in the Appendix, Table 8.

Machine steel (hot and cold rolled)	— soluble oil, lard oil
Tool steel (carbon and high-speed	— mineral lard oil — sulphur-base oil
Malleable iron	— soluble oil
Cast iron	— dry
Brass and bronze	— dry

Courtesy Kostel Enterprises Ltd.

Fig. 8-23
To correct the alignment of a tap, apply slight down pressure in the direction from which the tap leans while the tap is being turned.

THREADING DIES

Threading dies are used to cut external threads on round work. The most common threading dies are the adjustable and solid types. The *round adjustable die* (Fig. 8-24A) is split on one side and can be adjusted to cut slightly over or under-sized threads. It is mounted in a die stock, Fig. 8-25, which provides a means of turning the die on the work.

The *solid die*, Fig. 8-24B, cannot be adjusted and is generally used for recutting damaged or oversized threads. Solid dies are turned onto the thread with a standard open-end, twelve-point, or adjustable wrench.

Courtesy Cleveland Twist Drill (Canada) Ltd.

Fig. 8-24
Dies are used to cut external threads on a workpiece.

Courtesy Cleveland Twist Drill (Canada) Ltd.

Fig. 8-25
A die stock is used to turn the split die on the work.

To Thread with a Hand Die

1. Chamfer the end of the workpiece with a file or on the grinder.
2. Fasten the work securely in a vise. Hold small diameter work short to prevent if from bending.
3. Select the proper die and die stock.
4. Lubricate the tapered end of the die with a suitable cutting lubricant.
5. Place the tapered end of the die squarely on the work.
6. Press down on the die stock handles and turn clockwise several turns (Fig. 8-26).
7. Check the die to see that it has started squarely with the work.
8. If it is not square, remove the die from the work and restart it squarely, applying slight pressure while the die is being turned.
9. Turn the die forward one turn, and then reverse it approximately one-half of a turn to break the chip.
10. During the threading process apply cutting fluid frequently.
    CAUTION: When cutting a long thread, keep the arms and hands clear of the sharp threads coming through the die.

If the thread must be cut to a shoulder, remove the die and restart it with the tapered side of the die facing up. Complete the thread, being careful not to hit the shoulder; otherwise the work may be bent and the die broken.

Courtesy Kostel Enterprises Ltd.

Fig. 8-26
Start the tapered end of the die on the work.

HAND REAMING

A hand reamer (Fig. 8-27) is a cutting tool used for finishing drilled or bored holes to an accurate size and shape. The hand reamer is straight for nearly the full length of its flutes, but is tapered slightly on the end for a distance equal to its diameter in order to enable it to enter a hole.

Hand reamers are provided with a square at one end to fit a tap wrench and should never be used under mechanical power. A suitable cutting lubricant should be used, and not more than 0.12 mm (0.005 in.) should be removed with a hand reamer. Never turn a reamer backwards, or the cutting edges will be damaged.

Courtesy Cleveland Twist Drill (Canada) Ltd.

Fig. 8-27
A hand reamer is used to finish a drilled hole to shape and size.

Metric solid hand reamers are available in sizes from 1 to 13 mm in steps of 0.5 mm and from 13 to 26 mm in steps of 1 mm.

Inch size solid hand reamers are available in sizes from 1/8 to 1-1/2 in. in diameter in steps of 1/64 in.

METAL FASTENERS

In machine shop work, many methods are used to fasten work together. Some of the commoner methods, such as riveting, and fastening with screws and dowel pins, are briefly described.

RIVETING

A rivet is a metal pin made of soft steel, brass, copper, or aluminum, with a head at one end. The two most common types are the round head and the counter-sunk head (Fig. 8-28). Riveting consists of placing a rivet through holes in two or more pieces of metal and then forming a head on the other end of the rivet with a ball-peen hammer.

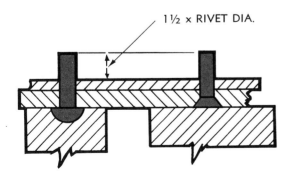

Fig. 8-28
Work set up for riveting.

To Rivet

1. Drill holes in the metal pieces 0.4 to 0.8 mm (1/64 to 1/32 in.) larger than the body of the rivet. If a countersunk rivet is being used, countersink for the head.
2. Insert a rivet through the holes, having it extend past the work 1-1/2 times the diameter of the rivet.
3. Place round head rivets in a metal block recessed for the shape of the head (Fig. 8-28). Countersunk rivets may be placed on a flat block.
4. Use a ball-peen hammer to form a head on the body of each rivet.

MACHINE SCREWS

Machine screws are widely used by the machinist for assembly work. A hole slightly larger than the body size of the screw may be drilled through the workpieces, the screw inserted, and a nut placed on the end of it. Another method of using machine screws is to drill a clearance hole through one piece and drill and tap the other piece to fit the thread of the screw. Some of the more common screws are shown in Fig. 8-29.

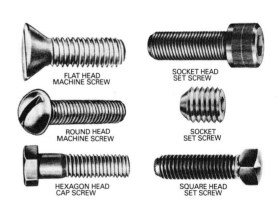

Courtesy Canadian Acme Screw and Gear Ltd. and Steel Company of Canada, Ltd.

Fig. 8-29
Threaded fasteners are made in a variety of types and sizes.

Work being fastened by flat head and socket head screws must be recessed so that the head of the screw is flush with the surface of the work. Flat head screws are recessed with an 82° countersink (Fig. 8-30). To recess work for socket head screws (Fig. 8-31), a counterbore or flat bottom drill is used.

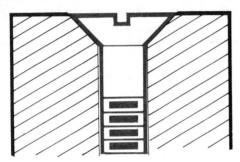

Fig. 8-30
The workpiece is countersunk properly when the head of the screw is flush with the work surface.

Courtesy Kostel Enterprises Ltd.

Fig. 8-31
Work must be counterbored when socket-head cap screws are used.

SELF-TAPPING SCREWS

Self-tapping screws (Fig. 8-32) are designed to cut a thread and therefore eliminate the need for tapping a hole. Self-tapping screws cut a mating thread in the work which fits snugly to the body of the screw. This close-fitting thread keeps the screw from backing off or coming loose, even under vibrating conditions. Self-tapping screws are used when assembling thin metal parts, nonferrous materials, and plastics.

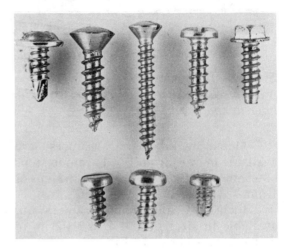

Courtesy Kostel Enterprises Ltd.

Fig. 8-32
A variety of self-tapping screws.

Courtesy H. Paulin & Co. Ltd.

A - Dowel pin

Courtesy Standco Canada Ltd.

B - Taper pin

Fig. 8-33
Dowel and taper pins are used to locate one part with another.

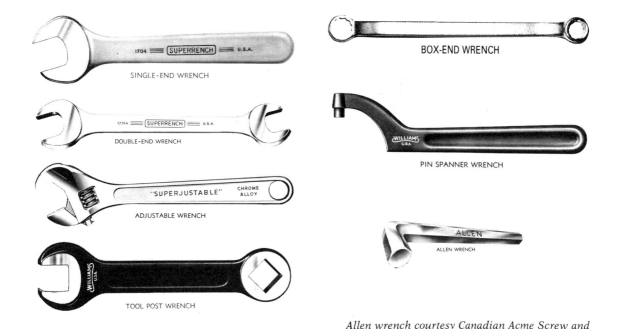

Allen wrench courtesy Canadian Acme Screw and Gear Ltd. Box wrench courtesy Kostel Enterprises Ltd. All others courtesy J.H. Williams & Co.

Fig. 8-34
Common wrenches used in machine shop work.

DOWEL AND TAPER PINS

Dowel pins (Fig. 8-33A) are used to locate one piece of work accurately with another. A hole is drilled and reamed through both workpieces simultaneously, and then the dowel pin is forced into the hole to keep both pieces in alignment. A tapered hole must be reamed to accommodate taper pins (Fig. 8-33B).

WRENCHES

Many types of wrenches are used in machine shop work, each being suited to a specific purpose. The name of a wrench is derived from either its shape, its use, or its construction. Various types of wrenches used in a machine shop are illustrated in Fig. 8-34.

A *single-end wrench* is one that fits only one size of bolt, head, or nut. The opening is generally offset at a 15° angle to permit complete rotation of a hexagonal nut in only 30° by "flopping" the wrench.

A *double-end wrench* has a different size opening at each end. It is used in the same manner as a single-end wrench.

The *adjustable wrench* is adjustable to various size nuts and is particularly useful for odd size nuts. Unfortunately this type of wrench, when not properly adjusted to the flats of the nut, will damage the corners of the nut. Adjustable wrenches should not be used on nuts which are very tight on the bolt. When excess pressure is applied to the handle, the jaws tend to spring, causing damage to the wrench and the corners of the nut. When using an adjustable wrench, the jaws should point in the direction of the force being applied (Fig. 8-35).

A *toolpost wrench* is a combination open-end wrench and a box-end wrench. The box end is used on toolpost screws and often on lathe carriage locking screws. In

WRENCHES 81

Fig. 8-35
The correct method of using an adjustable wrench.

order that the toolpost screw head will not be damaged, it is important that only this type of wrench be used.

Box-end or *twelve point wrenches*, are capable of operating in close quarters. The box-end wrench has twelve notches cut around the inside of the face. This type of wrench completely surrounds the nut and will not slip.

The *pin spanner wrench* fits around the circumference of a round nut. The pin on the wrench fits into a hole in the periphery of the nut.

The *socket set screw wrench*, commonly called the *Allen wrench*, is hexagonal and fits into the holes in safety set screws or socket-head set screws. They are made of tool steel in various sizes to suit the wide range of screw sizes. They are identified by the distance across the flats; this distance is usually one-half the outside diameter of the set screw thread in which it is used.

Using Wrenches

1. Always select a wrench which fits the nut or bolt properly.
   NOTE: *A wrench that is too large may slip off the nut and cause an accident.*
2. Whenever possible, *pull* rather than push on a wrench, in order to avoid injury if the wrench should slip.
3. Always be sure that the nut is fully seated in the wrench jaw.
4. Use a wrench in the same plane as the nut or bolt head.
5. When tightening or loosening a nut, a sharp, quick jerk is more effective than a steady pull.
6. A drop of oil on the threads, when assembling a bolt and nut, will ensure easier removal later.

SCREWDRIVERS

Screwdrivers (Fig. 8-36) are made in a variety of shapes, types, and sizes. The *standard* or *common* screwdriver is used on slotted-head screws. It consists of three parts: the *blade*, the *shank*, and the *handle*. Although most shanks are round, those on heavy-duty screwdrivers are generally square. This permits the use of a wrench to turn the screwdriver when extra torque is required.

The *offset* screwdriver is designed for use in confined areas where it is impossible to use a standard screwdriver. The blades on the ends are at right angles to each other. The screw is turned one-quarter of a turn with one end and then one-quarter of a turn with the other end.

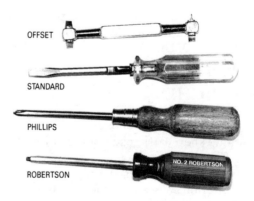

Fig. 8-36
Various types of commonly used screwdrivers.

Other commonly used screwdrivers are the *Robertson*, which has a square tip or blade, and the *Phillips*, which has a +-shaped point. Both types are made in different sizes to suit the wide range of screw sizes.

TEST YOUR KNOWLEDGE

The Machinist's Vise
1. How should the finished surface of a workpiece be protected when holding it in a vise?

Hammers
2. Describe the machinist's hammer.
3. Why are soft-faced hammers used?
4. Of what materials are soft-faced hammers made?
5. Why should a hammer be gripped at the end of the handle?
6. State three safety precautions that should be observed when using a hammer.

Metal Stamps
7. What should be done to cast iron or hot rolled steel before using metal stamps on it?
8. List the procedure for balancing stamping about a centre line.
9. Briefly describe the three-step method of using larger sized metal stamps.

Hand Hacksaw
10. In what direction should the teeth of a hacksaw blade point?
11. Of what materials are hacksaw blades manufactured?
12. Explain why two teeth of a blade should be in contact with the work at all times.
13. Briefly describe the procedure for mounting a blade on a hacksaw.
14. What pitch blade is recommended for:
    (a) general work?
    (b) tool steel?
    (c) angle iron and brass?
15. On what stroke is down-pressure applied when hacksawing?
16. How can thin pieces of metal be sawn?
17. Describe the procedure for starting a new blade in an old cut.

Chisels
18. Describe and state the purpose of:
    (a) a flat cold chisel
    (b) a round nose chisel
19. List three safety precautions to be observed while using a chisel.

Files
20. Name five parts of a file.
21. Of what material are files made?
22. Describe and state the purpose of:
    (a) single-cut files
    (b) double-cut files
23. Name the degrees of coarseness for larger files.
24. How is the degree of coarseness indicated on small files?

Using Files
25. Why should a file *never* be used without a handle?
26. On which stroke should down-pressure be applied to the file? Explain why.
27. Why should a hand never be rubbed across a surface being filed?
28. How can a file be prevented from clogging?

Filing Practice
29. Describe the procedure for filing a flat surface.
30. How can work be tested for flatness?
31. Describe the procedure for draw filing.

Hand Taps
32. Define a tap.
33. What is the purpose of the flutes on a tap?
34. Define the following information found on the shank of a tap:
    (a) 1/2 in. — 20 N.F.
    (b) M 30 × 3.5
35. Describe and state the purpose of:
    (a) taper tap
    (b) plug tap
    (c) bottoming tap

Tap Wrenches
36. Name two types of tap wrenches, and state where each is used.

Tap Drill Size
37. Define a tap drill.
38. What percentage of a thread will a tap cut after the hole has been drilled to tap drill size?
39. Using the tap drill formula, calculate the tap drill size for:
    (a) 7/16 in. — 14 N.C.
    (b) 3/4 in. — 10 N.C.
    (c) M 8 × 1.25
    (d) M 30 × 3.5

Tapping a Hole
40. Define the operation of tapping.
41. What care should be used while tapping a hole?
42. How should a tap be tested for squareness?
43. Explain the procedure for correcting a tap which has not started squarely.
44. Why should the tap be turned backwards after every quarter turn?
45. Describe the procedure for tapping a blind hole.

Tapping Lubricants
46. Why is a cutting lubricant used while tapping?
47. What lubricant is recommended for:
    (a) machine steel?
    (b) cast iron?

Threading Dies
48. For what purpose are threading dies used?
49. State the purpose of the adjustable die and the solid type die.
50. Explain the procedure for starting a die on work.
51. What procedure should be followed when it is necessary to thread up to a shoulder?

Hand Reaming
52. Describe and state the purpose of a hand reamer.
53. How much material should be removed with a hand reamer?

Metal Fasteners
54. Describe the procedure for riveting two pieces of metal together.
55. What tools are used to recess the work when using:
    (a) flat head screws?
    (b) socket head cap screws?
56. What is the purpose of dowel pins?

Wrenches
57. Name six types of wrenches used in a machine shop.
58. Why should an adjustable wrench not be used on an extremely tight nut?
59. Why is a box wrench a better type of wrench for most jobs?
60. Why is it advisable to pull rather than push a wrench?

Screwdrivers
61. Name three parts of a standard screwdriver.
62. What is the purpose of a square shank on a heavy-duty screwdriver?

CHAPTER 9
POWER SAWS

Courtesy DoAll Company

Archaeological discoveries show that the development of the first crude saw closely followed the origin of the stone axe and knife. The sharp edges of stones were serrated or toothed. This instrument cut by scraping away particles of the object being cut. A great improvement in the quality of saws followed the appearance of copper, bronze, and ferrous metals. With the modern steels and hardening methods, a wide variety of saw blades is available to suit hand hacksaws and machine power saws. Power saws fall into two general categories: cut-off saws and contour bandsaws.

CUT-OFF SAWS

Cut-off saws are used to rough-cut work to the required length from a longer piece of metal. The most common types of cut-off saws used in machine shop work are the power hacksaw, the horizontal bandsaw, the abrasive cut-off saw, and the circular cut-off saw. A brief description of each saw and its advantages follows.

The *power hacksaw*, which is a reciprocating type of saw, is usually permanently mounted to the floor. The saw frame and blade travel back and forth, with pressure being applied automatically only on the forward stroke. The power hacksaw finds limited use in machine shop work since it cuts only on the forward stroke, with resulting considerable wasted motion.

The *horizontal bandsaw* (Fig. 9-1) has a flexible, belt-like "endless" blade which cuts continuously in one direction. The thin, continuous blade travels over the rims of two pulley wheels and passes through roller guide brackets, which support the blade and keep it running true. Horizontal bandsaws are available in a wide variety of types and sizes, and are becoming increasingly popular because of their high production and versatility. On some models, casters are supplied with the machine so that it can be moved freely.

The *abrasive cut-off saw* (Fig. 9-2) uses a thin, abrasive grinding wheel, revolving at high speed, to cut off material. It can cut metals, glass, ceramics, etc. quickly and accurately, whether the material is hardened or not. This cut-off operation can be performed dry; however, cutting fluid is generally used to keep the work and saw cooler and to produce a better surface finish.

Courtesy Everett Industries Inc.

Fig. 9-2
The abrasive cut-off saw can cut hardened metals, glass, and ceramics.

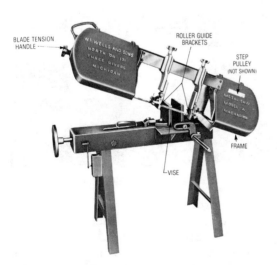

Courtesy Wells Manufacturing Corp.

Fig. 9-1
The main parts of a horizontal bandsaw.

The *cold circular cut-off saw* (Fig. 9-3) uses a circular saw blade similar to the one used on a wood-cutting table saw. The saw blade is generally made of chrome-

Courtesy Everett Industries Inc.

Fig. 9-3
A cold circular cut-off saw is used for cutting soft or unhardened metals.

vanadium steel; however, carbide-tipped blades are used for some cutting applications. These saws produce very accurate cuts on materials such as aluminum, brass, copper, machine steel, and stainless steel.

HORIZONTAL BANDSAW PARTS

The most common type of cut-off saw used in a school shop is the horizontal bandsaw, because it is easy to operate and cuts material quickly and accurately. The main operative parts of the horizontal bandsaw are:

The *saw frame*, hinged at the motor end, has two pulley wheels mounted on it, over which the continuous blade passes.

The *step pulleys* at the motor end are used to vary the speed of the continuous blade to suit the type of material being cut.

The *roller guide brackets* provide rigidity for a section of the blade and can be adjusted to accommodate various widths of material. These brackets should be adjusted to just clear the width of the work being cut.

The *blade tension handle* is used to adjust the tension on the saw blade. The blade should be adjusted to prevent it from wandering or twisting.

The *vise*, mounted on the table, can be adjusted to hold various sizes of workpieces. It can also be swivelled for making angular cuts on the end of a piece of material.

SAW BLADES

High-speed tungsten and high-speed molybdenum steel are commonly used in the manufacture of saw blades, and for the power hacksaw they are usually hardened completely. Flexible blades used on bandsaws have only the saw teeth hardened.

Saw blades are manufactured in various degrees of coarseness ranging from 4- to 14-pitch. When cutting large sections, use a coarse or 4-pitch blade, which provides the greatest chip clearance and helps to increase tooth penetration. For cutting tool steel and thin material, a 14-pitch blade is recommended. A 10-pitch blade is recommended for general purpose sawing. *Metric saw blades* are now available in similar sizes, but in teeth per 25 millimetres of length rather than teeth per inch. Therefore, the pitch of a blade having 10 teeth per 25 mm would be 10 ÷ 25 mm or 0.4 mm. Always select a saw blade as coarse as possible, but make sure that *two teeth* of the blade will be in contact with the work at all times. If less than two teeth are in contact with the work, the work can be caught in the tooth space (gullet), which would cause the teeth of the blade to strip or break.

INSTALLING A BLADE

When replacing a blade, always make sure that the teeth are pointing in the direction of saw travel or toward the motor end of the machine. The blade tension should be adjusted to prevent the blade from twisting or wandering during a cut. If it is necessary to replace a blade before a cut is finished, rotate the work one-quarter of a turn in the vise. This will prevent the new blade from jamming or breaking in the cut made by the worn saw.

To Install a Saw Blade

1. Loosen the blade tension handle.
2. Move the adjustable pulley wheel forward slightly.
3. Mount the new saw band over the two pulleys.
   NOTE: *Be sure that the saw teeth are pointing toward the motor end of the machine.*
4. Place the saw blade between the rollers of the guide brackets (Fig. 9-4).
5. Tighten the blade tension handle only enough to hold the blade on the pulleys.
6. Start and quickly stop the machine in order to make the saw blade revolve a turn or two. This will seat the blade on the pulleys.
7. Tighten the blade tension handle as tightly as possible with *one hand*.

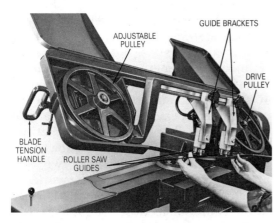

Fig. 9-4
Installing a blade on a horizontal bandsaw.

SAWING

For the most efficient sawing, it is important that the correct type and pitch of saw blade be selected and that it be run at the proper speed for the material being cut. Use finer-tooth blades when cutting thin cross-sections and extra-hard materials. Coarser-tooth blades should be used for thick cross-sections and material which is soft and stringy. The blade speed should suit the type and thickness of the material being cut. Too fast a blade speed or excessive feeding pressure will dull the saw teeth quickly and cause an inaccurate cut. Too slow a blade speed does not use the saw efficiently, wastes time, and results in lower production.

To Saw Work to Length

1. Check the solid vise jaw with a square to make sure it is at right angles to the saw blade.
2. Place the material in the vise, supporting long pieces with a floor stand (Fig. 9-5).

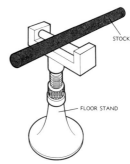

Fig. 9-5
A floor stand is used to support long workpieces held in a cut-off saw.

3. Lower the saw blade until it just clears the work. Keep it in this position by engaging the ratchet lever or by closing the hydraulic valve.
4. Adjust the roller guide brackets until they *just clear* both sides of the material to be cut.
5. Hold a steel rule against the edge of the saw blade and move the material until the correct length is obtained.
6. Always allow 1.5 mm (1/16 in.) for each 25 mm (1 in.) of thickness longer than required to compensate for any *saw run-out* (slightly angular cut caused by hard spots in steel or a dull saw blade).

Courtesy Wells Manufacturing Corp.

Fig. 9-6
The stop gauge is used to control the length of the cut.

7. Tighten the vise and recheck the length from the blade to the end of the material to make sure the the work has not moved.
8. Raise the saw frame slightly, release the ratchet lever or open the hydraulic valve, and then start the machine.
9. Lower the blade slowly until it just touches the work.
10. When the cut has been completed, the machine will shut off automatically.

Sawing Hints
1. Never attempt to mount, measure, or remove work unless the saw is stopped.
2. Guard long material at both ends to prevent anyone from coming in contact with it.
3. Use cutting fluid, whenever possible, to help prolong the life of the saw blade.
4. When sawing thin pieces, hold the material flat in the vise to prevent the saw teeth from breaking.

5. Caution should be used when applying extra force to the saw frame, as this generally causes work to be cut out of square.
6. When several pieces of the same length are required, set the stop gauge which is supplied with most cut-off saws (Fig. 9-6).
7. When holding short work in a vise, be sure to place a short piece of the same thickness in the opposite end of the vise. This will prevent the vise from twisting when it is tightened (Fig. 9-7).

Courtesy Kostel Enterprises Ltd.

Fig. 9-7
A spacer block should be used when clamping short work in a vise.

CONTOUR-CUTTING BANDSAW

The contour-cutting (vertical) bandsaw has been widely accepted by industry as a fast and economical method of cutting metal and other materials since its development in the early 1930s. It has provided industry with a fast and accurate method of sawing, filing, and polishing straight and contour shapes.

CONTOUR-CUTTING BANDSAW 89

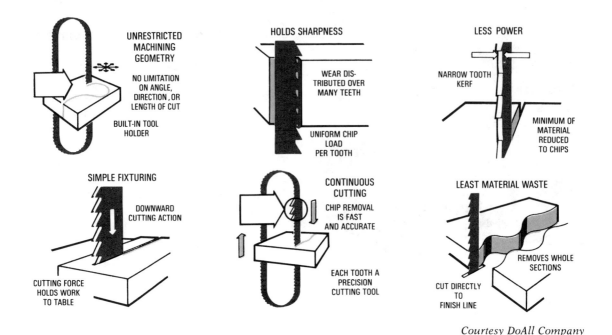

Fig. 9-8
Advantages of the contour-cutting bandsaw.

The contour-cutting bandsaw has a flexible "endless" blade, which cuts continuously in one direction. This blade is fitted around two vertically mounted pulleys with flat rubber-tired rims on which the blade rides. Saw guides immediately above and below the table of the saw help to guide and support the vertical saw blade.

The contour-cutting bandsaw has many advantages in the metal-cutting trade and some of the more common ones are illustrated in Fig. 9-8.

CONTOUR BANDSAW PARTS

Although contour bandsaws are available in a wide variety of sizes and shapes, each machine contains some parts which are basic to all machines (Fig. 9-9).

The *job selector* lists the cutting speeds for various materials and pitches of blade.

The *upper pulley* is used to support and tension the saw band. The tension is adjusted by the *tension handwheel*.

The *lower pulley* is driven by a variable-speed pulley and can be adjusted to various speeds by the *speed change handwheel*.

The *upper and lower saw guides* help support the blade and prevent it from wandering.

The *table* can be tilted up to 45° for the cutting of angular surfaces.

The *butt welder*, supplied with most machines, is used to weld and anneal saw bands.

TYPES OF SAW BLADES

Saw blades used on contour bandsaws are generally made of carbon-alloy steel, high-speed steel, and tungsten-carbide tipped blades. These blades are available in three types of tooth forms (Fig. 9-10A), and each is available in three types of sets (Fig. 9-10B).

The *precision or regular tooth* blade is a general purpose blade, used where a fine finish and accurate cut are required.

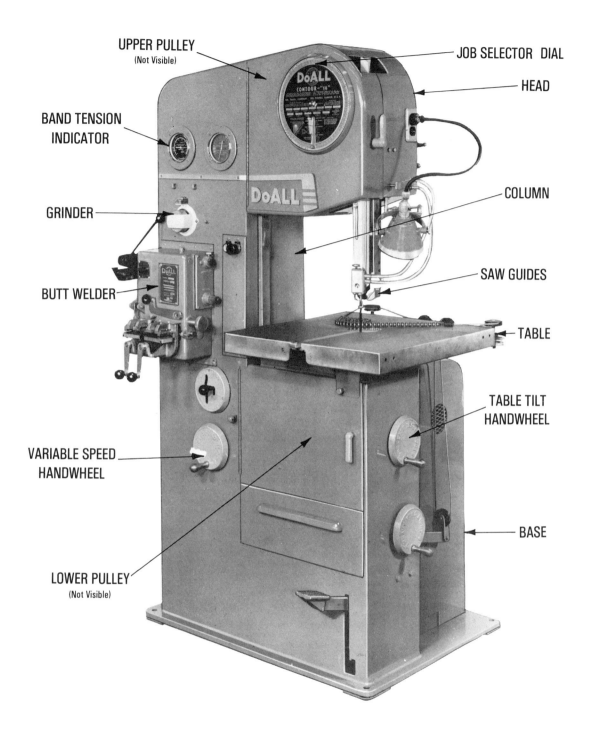

Fig. 9-9
A contour bandsaw.

Courtesy DoAll Company

Fig. 9-10A
Saw tooth forms.

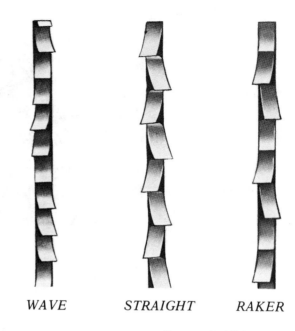

WAVE   STRAIGHT   RAKER

Courtesy DoAll Company

Fig. 9-10B
The common tooth sets found on saw blades.

The *claw or hook tooth* blade allows rapid removal of chips without increasing friction heat. It is used for sawing wood, plastics, nonferrous metals, and large sections of ferrous castings and machine steel. The *buttress or skip tooth* blade has wide tooth spacing for good chip clearance and high-speed cutting of nonferrous metals, wood, plastics, etc.

On large work sections, a coarse pitch blade should be used. For thin sections, select a fine pitch blade. As in all other methods of sawing, a general rule to follow is that not less than two teeth should be in contact with the work at all times.

BLADE LENGTH

Metal-cutting saw bands are usually packaged in coils about 30 to 150 m (100 to 500 ft.) in length. The length required is cut from the coil, and its two ends are then welded together.

To calculate the length required for a two-wheel bandsaw, take twice the centre distance between each pulley and to it add the circumference of one wheel. This is the total length of the saw band.

When measuring the centre distance between wheels, make sure the top wheel is lowered approximately 25 mm (1 in.) from its top position. This will allow for stretching when the blade is tensioned.

BANDSAW OPERATIONS

With the proper attachments, a wide variety of operations can be performed on a contour bandsaw. The procedure for performing some of the more common operations is listed in the following section.

To Weld a Saw Band

Most contour bandsaws contain an attachment or accessory for welding a saw band to make a continuous loop. Since the operation of these welding units varies from machine to machine, follow the instructions in the operator's manual when welding a saw band.

To Mount a Saw Blade

1. Select the correct saw guides for the width of blade being used.
2. Mount both upper and lower saw guides on the machine (Fig. 9-11).
3. Adjust the upper saw guide until it clears the top of the work by about 6 mm (1/4 in.)
4. Insert the saw plate in the table.
5. Place the saw band on both upper and lower pulleys, making sure that the teeth are pointed in the direction of band travel (toward the table).
6. Adjust the upper pulley with the *tension handwheel* until some tension is registered on the tension gauge.
7. Set the gearshift lever to neutral and turn the upper pulley by hand to see that the saw blade is tracking properly on the pulley.
8. Re-engage the gearshift lever and start the machine.
9. Adjust the blade to the recommended tension as indicated on the tension gauge by using the tension handwheel.
10. Check that the back of the blade just touches the back of the saw guides. If the blade is not tracking properly, use the handwheel at the rear of the frame (at the opposite side to the speed change handwheel) to tilt the upper pulley.

To Saw to a Layout

1. Set the machine to the proper speed for the type of blade and the material being cut. Consult the job selector (Fig. 9-9).
2. Use the work-holding jaw (Fig. 9-12) or a piece of wood to feed the work into the saw.

NOTE: Keep the fingers well clear of the moving saw blade.

Courtesy DoAll Company

Fig. 9-12
The work-holding jaw is used to feed the work into the saw.

3. Feed the work into the saw blade steadily. Do not apply too much pressure and crowd the blade.
4. Cut to within approximately 0.8 mm (1/32 in.) of the layout lines. This allows material for finishing.
5. Never attempt to cut too small a radius with a wide saw. This will damage the blade and the lower drive wheel.
6. The table may be tilted up to 45° for cutting angular surfaces (Fig. 9-13).

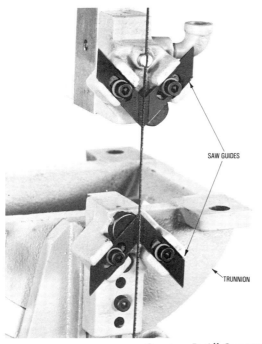

Courtesy DoAll Company

Fig. 9-11
Upper and lower saw guides are used to guide and support the saw band.

Courtesy DoAll Company

Fig. 9-13
This composite photograph shows the table in a horizontal position and tilted 45°.

To Mount a File Band
1. Remove the saw blade, upper and lower saw guides, and the saw plate.
2. Mount the correct file guide support on the lower post block (Fig. 9-14).
3. Insert the file plate into the table.
4. Mount the file guide to the upper post.
5. Lower the upper post until the file guide is below the centre of the file guide support.
6. Thread the file band upward through the hole in the table centre disc.
   NOTE: *Make sure the unhinged end of the file segment is pointing upward.*
7. Join the ends of the file band (Fig. 9-15), and then apply the proper amount of tension to the file.
8. If necessary, adjust the upper pulley to make the file band track freely in the file guide channel.

To File on a Contour Bandsaw
1. Consult the job selector (Fig. 9-9) and set the machine to the proper speed. The best filing speeds are between 15 and 30 m (50 and 100 ft.) per minute.
2. Apply light work pressure to the file band.

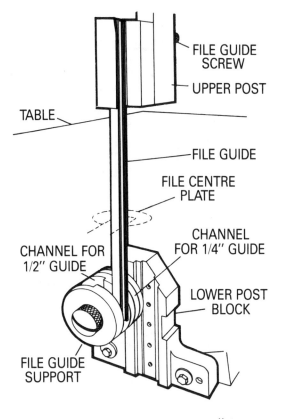

Courtesy DoAll Company

Fig. 9-14
A file guide and support mounted on a contour bandsaw.

It not only gives a better finish, but prevents the teeth from becoming clogged.
3. Keep moving the work sideways against the file to prevent filing grooves in the work.
4. Use a file card to keep the file clean. Loaded files cause bumpy filing and scratches in the work.
   NOTE: *Stop the machine before attempting to clean the file.*

TEST YOUR KNOWLEDGE
Cut-off Saws
1. Name and describe the cutting action of two types of cut-off saws.

Courtesy DoAll Company

Fig. 9-15
Mounting a file band on a contour bandsaw.

Horizontal Bandsaw Parts
2. What is the purpose of the roller guide brackets?
3. How tight should the blade tension handle be adjusted?

Saw Blades
4. Of what material are saw blades made?
5. What pitch blade is recommended for:
   (a) large sections?
   (b) thin sections?
   (c) general purpose sawing?
6. Why should two teeth of a saw blade be in contact with the work at all times?

Installing a Blade
7. In what direction should the teeth of a saw blade point?
8. How should the workpiece be set when sawing partially cut work with a new saw blade?

Sawing
9. What may happen if too fast a blade speed is used?
10. How can a vise be checked for squareness?
11. How should the roller guide brackets be set when cutting work?
12. Explain how thin work should be held in the vise.
13. What precaution should be used when holding short work in a vise?

Contour-Cutting Bandsaw
14. List six advantages of the contour bandsaw.
15. Explain the purpose of the following:
    (a) upper pulley
    (b) lower pulley
    (c) saw guides

Types of Saw Blades
16. Name and state the purpose of three types of saw blades.
17. What general rule should be followed when selecting the pitch of a saw blade?

Mounting a Saw Blade
18. In what position should the upper saw guide be set?
19. In what direction should the teeth of a saw blade point?
20. How can the upper pulley be adjusted if the saw is not tracking properly?

Sawing to a Layout
21. How should work be fed into a saw blade?
22. What would be the result of trying to cut a small radius with a wide blade?
23. How can angular surfaces be cut on a contour bandsaw?

Filing
24. What accessories must be mounted on the bandsaw for filing?
25. List the procedure for mounting a file band on the bandsaw.
26. What speeds are recommended for filing?
27. Why should only light work pressure be applied to the file band?
28. How can a loaded or clogged file be cleaned?

TEST YOUR KNOWLEDGE

CHAPTER 10
DRILL PRESSES

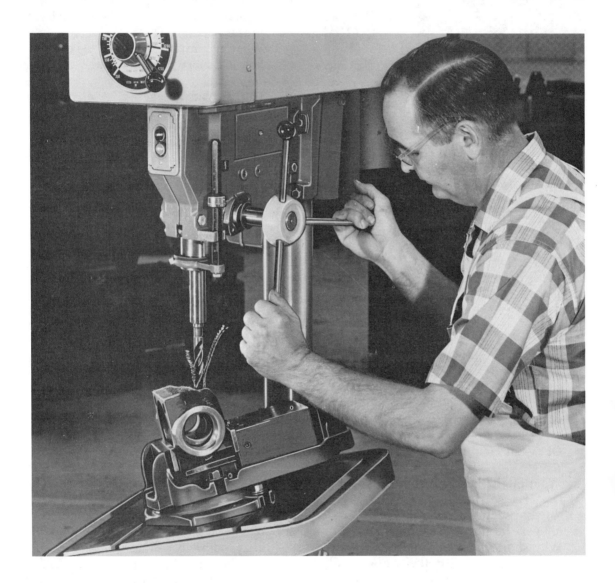

The drill press, probably the first mechanical device developed in prehistoric times, is one of the most commonly used machines in a machine shop. The main purpose of a drill press is to grip, revolve, and feed a twist drill in order to produce a hole in a piece of metal or other material. Main parts of any drill press include the *spindle*, which holds and revolves the cutting tool, and the *table* upon which the work is held or fastened. The revolving drill or cutting tool is generally fed into the workpiece manually on bench type drill presses, and either manually or automatically on floor type drill presses. The variety of cutting tools and attachments which are available allow operations such as drilling, reaming, countersinking, counterboring, tapping, spot facing, and boring to be performed (Fig. 10-1).

DRILL PRESS TYPES AND CONSTRUCTION

Drill presses are available in a wide variety of types and sizes to suit industry. These range from the small hobby-type drill press to the larger, more complex, and numerically controlled machines used by industry for production purposes. The most common machines found in a machine shop are the *bench type sensitive drill press* and the *floor type drill press*. Other drill presses, such as the upright, post, radial, horizontal, gang, portable, multiple spindle, and numerically controlled types, are variations of the standard machine and are generally designed for specific purposes.

Drill presses may be purchased in a wide variety of sizes to suit the purpose for which they will be used. The size of any drill press is generally given as the distance from the *edge of the column* to the centre of the *drill press spindle*. For example, on a 250 mm drill press it would be possible to drill a hole in a workpiece 250 mm from its edge; on a 10 in. drill press a hole could be drilled 10 in. from the edge of a workpiece. The vertical capacity of a drill press is measured when the head is in the highest position and the table in its lowest position.

DRILL PRESS PARTS

Although drill presses are manufactured in a wide variety of types and sizes, all

A - Drilling   B - Reaming

 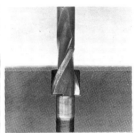

C - Countersinking   D - Counterboring

E - Tapping   F - Boring

Courtesy Kostel Enterprises Ltd.

Fig. 10-1
Common operations performed on a drill press.

drilling machines contain certain basic parts. The main parts on the bench and floor type models are *base, column, table,* and *drilling head* (Fig. 10-2). The floor type model is larger and has a longer column than the bench type.

BASE — The base, usually made of cast iron, provides stability for the machine and rigid mounting for the column. The base is usually provided with holes so that

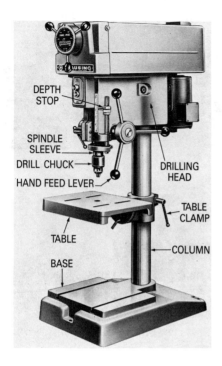

Courtesy Clausing Corp.

Fig. 10-2
The main parts of a sensitive drill press.

DRILLING HEAD — The head, mounted close to the top of the column, contains the mechanism which is used to revolve the cutting tool and advance it into the workpiece. The *spindle*, which is a round shaft that holds and drives the cutting tool, is housed in the *spindle sleeve* or *quill*. The spindle sleeve does not revolve, but slides up and down inside the head to provide a downfeed for the cutting tool. The end of the spindle may have a tapered hole to hold taper shank tools, or it may be threaded or tapered for attaching a *drill chuck* (Fig. 10-2).

The *hand feed lever* is used to control the vertical movement of the spindle sleeve and the cutting tool. A *depth stop*, attached to the spindle sleeve, can be set to control the depth that a cutting tool enters the workpiece.

it may be bolted to a table or bench. The slots or ribs in the base allow the work-holding device or the workpiece to be fastened to the base.

COLUMN — The column is an accurate cylindrical post which fits into the base. The table, which is fitted on the column, may be adjusted to any point between the base and head. The head of the drill press is mounted near the top of the column.

TABLE — The table, either round or rectangular in shape, is used to support the workpiece to be machined. The table, whose surface is at 90° to the column, may be raised, lowered, and swivelled around the column. On some models it is possible to tilt the table in either direction for drilling holes on an angle. Slots are provided in most tables to allow jigs, fixtures, or large workpieces to be clamped directly to the table.

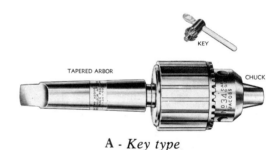

A - *Key type*

B - *Keyless*

Courtesy Jacobs Manufacturing Co.

Fig. 10-3
Drill chucks are used to grip straight shank cutting tools.

TOOL-HOLDING DEVICES

The drill press spindle provides a means of holding and driving the cutting tool. It may have a tapered hole to accommodate taper shank tools, or its end may be tapered or threaded for mounting a drill chuck. Although there is a variety of drill press tool-holding devices and accessories, the most common in a machine shop are drill chucks, drill sleeves, and drill sockets.

DRILL CHUCKS

Drill chucks are the most common devices used on a drill press for holding straight shank cutting tools. Most drill chucks contain three jaws that move simultaneously (all at the same time) when the outer sleeve is turned or, on some types of chucks, when the outer collar is raised. The three jaws hold the straight shank of a cutting tool securely and cause it to run accurately. There are two common types of drill chucks: the key type and the keyless type.

The *key-type chuck* (Fig. 10-3A) may be provided with a tapered arbor, which fits into the tapered hole of the spindle, or it may have a tapered or threaded hole for fastening to the end of the drill press spindle. A *key* is used to turn the outer collar, which causes the jaws to grip a straight shank cutting tool.

The *keyless drill chuck* (Fig. 10-3B) is generally used in production work because some models allow cutting tools to be inserted and removed while the machine is in operation. Two common keyless chucks are the Albrecht Precision and the Jacobs Impact chucks.

DRILL SLEEVES AND SOCKETS

The size of the tapered hole in the drill press spindle is generally in proportion to the size of the machine: the larger the machine, the larger the spindle hole. The size of the tapered shank on cutting tools is also manufactured in proportion to the

A - Sleeve   B - Socket

Courtesy Kostel Enterprises Ltd.

Fig. 10-4
Drill sleeves and sockets are used to adapt the tapered shank of cutting tools to the drill press spindle.

size of the tool. *Drill sleeves* (Fig. 10-4A) are used to adapt the cutting tool shank to the machine spindle if the taper on the cutting tool is smaller than the tapered hole in the spindle.

A *drill socket* (Fig. 10-4B) is used when the hole in the spindle of the drill press is too small for the taper shank of the drill. The drill is first mounted in the socket, and then the socket is inserted into the drill press spindle. Drill sockets may also be used as extension sockets to provide extra length.

MOUNTING AND REMOVING TAPER SHANK TOOLS

Before a taper shank tool is mounted in a drill press spindle, be sure that the external taper of the tool and the internal taper of the spindle are thoroughly cleaned. Align the tang of the tool with the slot in the spindle hole, and with a sharp snap, force the tool into the spindle.

A *drift*, a wedge-shaped tool, is used to remove a taper shank tool from the drill press spindle (Fig. 10-5). The drift should be inserted into the spindle slot with its *rounded edge up*. Place a wooden block on the drill press table to prevent the drill from marring the table when it is removed. Sharply strike the end of the drift with a hammer to remove the tool from the drill press spindle.

Courtesy Cleveland Twist Drill (Canada) Ltd. and Kostel Enterprises Ltd.

Fig. 10-5
The wooden block prevents damage to the table if the drill should fall when it is being removed from the spindle.

TWIST DRILLS

A *twist drill* is an end-cutting tool used to produce a hole in a piece of metal or other material. The most common drill manufactured has two cutting edges (lips) and two straight or helical flutes, which provide the cutting edges, admit cutting fluid, and allow the chips to escape during the drilling operation.

The most common twist drills used in a machine shop are made of high-speed steel and cemented carbides. A recent development is the coating of regular drills with titanium carbide to improve their performance.

High-speed steel drills are the most commonly used drills, since they can be operated at good speeds and the cutting edges can withstand heat and wear.

Cemented carbide drills, which can be operated much faster than high-speed steel drills, are used to drill hard materials. Cemented carbide drills have found wide use in production work because they can be operated at high speeds, the cutting edges do not wear rapidly, and they are capable of withstanding higher heat.

Titanium nitride coated drills can be run at higher speeds and feeds than regular high-speed drills. They outperform regular drills by 7 to 10 times, while at the same time producing holes with a better surface finish.

TWIST DRILL PARTS

A twist drill (Fig. 10-6) may be divided into three main sections: the *shank, body,* and *point*.

SHANK — The shank is the part of the drill that fits into a holding device that revolves the drill. The shanks of twist drills may be either straight or tapered (Fig. 10-6). Straight shanks are generally provided on drills up to 12 mm (or 1/2 in. for inch drills) in diameter, while drills over 12 mm (or 1/2 in.) diameter usually have tapered shanks. Straight shank drills are held in some type of drill chuck, while taper shank drills fit into the internal taper of the drill press spindle.

Courtesy Cleveland Twist Drill (Canada) Ltd.

Fig. 10-6
The main parts of a twist drill.

The *tang*, at the small end of the tapered shank, is machined flat to fit the slot in the drill press spindle. Its main purpose is to allow the drill to be removed from the spindle with a drift without damaging the shank. The tang may also prevent the shank from turning in the drill press spindle because of a poor fit on the taper or too much drilling pressure.

BODY — The body of a twist drill consists of that portion between the shank and the point. The body contains the *flutes, margin, body clearance,* and *web* of the drill.

(d) The *web* is the thin metal partition in the centre of the drill, which extends the full length of the flutes. This part forms the chisel edge at the cutting end of the drill. The web gradually increases in thickness toward the shank to give the drill strength (Fig. 10-7).

POINT — The point of a twist drill consists of the entire cone-shaped cutting end of the drill. The shape and condition of the point are very important to the cutting action of the drill. The drill point consists of the *chisel edge, cutting edges or lips, lip clearance*, and *heel* (Fig. 10-8).

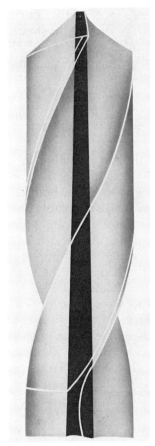

Courtesy Cleveland Twist Drill (Canada) Ltd.

Fig. 10-7
The thickness of the web increases toward the shank to strengthen the drill.

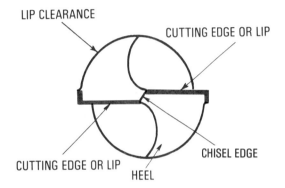

Fig. 10-8
The parts of a twist drill point.

(a) The *flutes* on most drills consist of two or more helical grooves cut along the body of the drill. The flutes form the cutting edges of the drill, provide them with rake, admit cutting fluid, and allow chips to escape during the drilling operation.

(b) The *margin* is the narrow, raised section on the body immediately next to the flutes. The diameter of the drill is measured across the margin, which extends the full length of the flutes.

(c) The *body clearance* is the undercut portion of the body between the margin and the flute.

(a) The *chisel edge* is that portion which connects the two cutting edges. It is formed by the intersection of the cone-shaped surface of the point. The cutting action of the chisel edge is not very good; when drilling holes over about 12 mm (1/2 in.) in diameter, it is wise to first drill a lead or pilot hole in the workpiece to relieve some of the pressure on the drill point.

(b) The *cutting edges or lips* are formed by the intersection of the flutes and the cone-shaped point. The lips must both be the same length and have the same angle, so that the drill will run true and not cut a hole larger than the size of the drill.

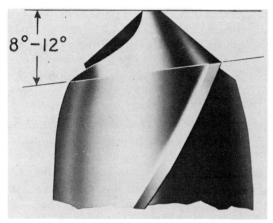

Courtesy Cleveland Twist Drill (Canada) Ltd.

Fig. 10-9
The average lip clearance behind the cutting edges ranges from 8 to 12°.

(c) The *lip clearance* is the relief which is ground on the point of the drill extending from the cutting lips back to the heel (Fig. 10-9). Lip clearance allows the lips of the drill to cut into the metal without the heel rubbing. The average lip clearance is from 8 to 12°, depending on the type of the material to be drilled.

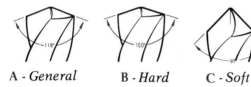

A - *General purpose*   B - *Hard*   C - *Soft*

Fig. 10-10
Drill point angles for various materials.

DRILL POINT ANGLES AND CLEARANCES

For general purpose drilling, the drill point should be ground to an included angle of 118° (Fig. 10-10A), and the lip clearance should range from 8 to 12°. The drill point for hard materials should be ground to an included angle from 135 to 150° (Fig. 10-10B), and the lip clearance should be from 8 to 10°. For drilling soft materials, the drill point should be ground to an included angle of 90° (Fig. 10-10C), with the lip clearance ranging from 15 to 18°.

SYSTEMS OF DRILL SIZES

Twist drills are available in both metric and inch sizes. Inch drills are designated by *fractional*, *number*, and *letter* systems. *Metric* drills are available in various set ranges. The size of straight shank drills is marked on the shank, while taper shank drills are generally stamped on the neck between the body and the shank.

(a) *Fractional* inch drills are manufactured in sizes from 1/64 to 3-1/2 in. in diameter, varying in steps of 1/64 in. from one size to the next. Drills larger than 3-1/2 in. in diameter must be ordered specially from the manufacturer.

(b) *Number* size drills range from the #1 drill (0.228 in.) to the #97 drill (0.0059 in.). The most common number drill set contains drills from #1 to #60. The large range of sizes enables almost any hole between 0.0059 to 0.228 in. to be drilled.

(c) *Letter* size drills range from A to Z. The letter A drill is the smallest in the set (0.234 in.) and the letter Z is the largest (0.413 in.).

(d) *Metric* size drills are available in various sets, but are not designated by various systems. The miniature metric drill set ranges from 0.04 to 0.99 mm in steps of 0.01 mm. Straight shank metric drills are available in sizes from 0.5 to 20 mm, ranging in steps of 0.02 to 1 mm, depending on the size. Taper shank metric drills are available in sizes from 8 to 80 mm.

See the tables in the Appendix for the number, letter, and metric drill size tables.

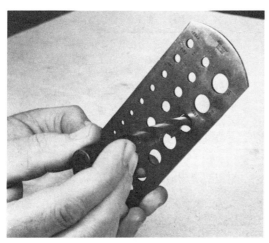

Fig. 10-11A
Checking the size of a drill with a drill gauge.

Fig. 10-11B
Checking the size of a drill with a micrometer.

MEASURING THE SIZE OF A DRILL

In order to produce a hole to the required size, it is important that the correct size drill is used to drill the hole. It is good practice to always check a drill for size before using it to drill a hole. Drills may be checked for size by two methods: with a drill gauge (Fig. 10-11A) and with a micrometer (Fig. 10-11B). When a drill is being checked for size with a micrometer, always be sure that the measurement is taken across the margin of the drill.

CUTTING SPEEDS AND FEEDS

The selection of the proper speeds and feeds for the cutting tool to be used and the type of material being drilled are important factors which the operator must consider. These two factors affect the amount of time required to complete an operation (production rate) and how long a cutting tool will perform satisfactorily. Time will be wasted unnecessarily if the speed and feed are too low, while the cutting tool will wear quickly if the speed and feed are set too high.

CUTTING SPEED

The speed at which a twist drill should be operated is often referred to as *cutting speed*, surface speed, or peripheral speed. Cutting speed may be defined as the distance, in either surface metres or surface feet, that a point on the circumference of the drill travels in one minute. For example, if tool steel has a recommended cutting speed of 30 m (100 ft.) per minute, the drill press speed should be set so that a point on the circumference of the drill will travel 30 m (100 ft.) in one minute. The wide range of drill sizes used for drilling holes in the various kinds of metals requires an equally wide range of speeds at which the drills can be efficiently operated. The size of the drill, the material it is made from, and the type of material to be drilled must all be taken into account when determining a safe and efficient speed at which to operate a drill press.

As a result of many years of research, cutting tool and steel manufacturers recommend that various types of metal be machined at certain cutting speeds for the best production rates and the best tool life.

TABLE 10-1 CUTTING SPEEDS FOR HIGH-SPEED STEEL DRILLS										
		Steel Casting		Tool Steel		Cast Iron		Machine Steel		Brass and Aluminum	
		Cutting Speeds									
		m/min.	ft./min.	m/min.	ft./min.	m/min.	ft./min.	m/min.	ft./min.	m/min.	ft./min.
Size		12	40	18	60	24	80	30	100	60	200
milli-metre	inch	Revolutions Per Minute									
2	1/16	1910	2445	2865	3665	3820	4890	4775	6110	9550	12225
3	1/8	1275	1220	1910	1835	2545	2445	3185	3055	6365	6110
4	3/16	955	815	1430	1220	1910	1630	2385	2035	4775	4075
5	1/4	765	610	1145	915	1530	1220	1910	1530	3820	3055
6	5/16	635	490	955	735	1275	980	1590	1220	3180	2445
7	3/8	545	405	820	610	1090	815	1305	1020	2730	2035
8	7/16	475	350	715	525	955	700	1195	875	2390	1745
9	1/2	425	305	635	460	850	610	1060	765	2120	1530
10	5/8	350	245	520	365	695	490	870	610	1735	1220
15	3/4	255	205	380	305	510	405	635	510	1275	1020
20	7/8	190	175	285	260	380	350	475	435	955	875
25	1	150	155	230	230	305	305	380	380	765	765

The recommended cutting speeds for various materials are listed in Table 10-1.

Whenever reference is made to the *speed* of a drill, the cutting speed in *surface metres or in surface feet per minute* is implied, and not in revolutions per minute (*r/min*), unless specifically stated.

REVOLUTIONS PER MINUTE

The number of revolutions necessary to produce the desired cutting speed is called *r/min* (revolutions per minute). A small drill operating at the same *r/min* as a larger drill will travel fewer metres or feet per minute and naturally will cut more efficiently at a higher number of *r/min*.

To find the number of revolutions per minute at which a drill press spindle must be set to obtain a certain cutting speed, the following information must be known: (a) the recommended cutting speed of the material to be drilled, (b) the type of material from which the drill is made, and (c) the diameter of the drill. To calculate the spindle speed for any machine, divide the cutting speed of the metal by the circumference of the rotating member, which may be a drill, a milling cutter, or the workpiece in a lathe.

Apply one of the following formulas to calculate the spindle speed (*r/min*) at which the drill press should be set. (Table 10-1 lists the cutting speeds for various materials.)

Inch Drills

$$r/min = \frac{CS \times 12}{\pi \times D}$$

CS = cutting speed of the material in feet per minute.
D = diameter of the drill in inches.

Since only a few machines are equipped with variable speed drives, which allow them to be set to the exact calculated speed, a simplified formula can be used to calculate *r/min*. The π (3.1416) on the bottom line of the formula will divide into 12 of the top line approximately four times. This results in a simplified formula which is close enough for most drill presses.

$$r/min = \frac{CS \times 4}{D}$$

EXAMPLE: Calculate the *r/min* at which the drill press should be set to drill a 1/2 in. diameter hole in a piece of machine steel. (See Table 10-1 for the cutting speed of machine steel.)

$$r/min = \frac{CS \times 4}{D}$$
$$= \frac{100 \times 4}{1/2}$$
$$= 800$$

When it is not possible to set the drill press to the exact speed, always set it to the closest speed *under* the calculated speed.

Metric Drills

$$r/min = \frac{CS \text{ (metres)}}{\pi \times D \text{ (millimetres)}}$$

Since the cutting speed is usually given in metres and the diameter of a workpiece is expressed in millimetres, it is necessary to convert the metres in the numerator to millimetres so that both parts of the equation are in the same unit. Therefore multiply the *CS* in metres by 1000 to bring it to millimetres.

$$r/min = \frac{CS \times 1000}{\pi \times D}$$

EXAMPLE: Calculate the *r/min* at which a drill press should be set to drill a 12 mm hole in a piece of machine steel.

$$r/min = \frac{30 \times 1000}{3.1416 \times 12}$$
$$= \frac{30000}{37.699}$$
$$= 796$$

Since only a few machines are equipped with a variable speed drive, a simplified formula, which is suitable for most drilling operations, can be derived by dividing π (3.1416) into 1000:

$$r/min = \frac{CS \times 1000}{\pi \times D}$$
$$= \frac{CS \times 320}{D}$$

The same drilling problem, solved by using the simplified formula, is as follows:

$$r/min = \frac{30 \times 320}{12}$$
$$= \frac{9600}{12}$$
$$= 800$$

FACTORS AFFECTING DRILL SPEED (r/min)

The calculated drill speed may have to be varied slightly to suit the following factors:
(a) the type and condition of the machine;
(b) the accuracy and finish of the hole required;
(c) the rigidity of the work setup;
(d) the use of cutting fluid.

FEED

Feed is the distance that a drill advances into the work for each complete revolution. The feed rate is important because it affects both the life of the drill and the rate of production. Too coarse a feed may cause the cutting edges to break or chip, while too fine a feed causes a drill to chatter, which dulls the cutting edges. The recommended feeds per revolution for millimetre and fractional inch size drills are listed in Table 10-2.

TABLE 10-2 RECOMMENDED DRILL FEEDS			
Drill Size		Feed per Revolution	
Millimetre	Inch	Millimetres	Inches
3 or less	1/8 or less	0.02—0.05	0.001—0.002
3—6	1/8—1/4	0.05—0.10	0.002—0.004
6—12	1/4—1/2	0.10—0.17	0.004—0.007
12—15	1/2—1	0.17—0.37	0.007—0.015
25—38	1—1-1/2	0.37—0.63	0.015—0.025

TABLE 10-3 CUTTING LUBRICANTS FOR DRILL PRESS WORK			
Material	Drilling	Reaming	Tapping
Machine steel (hot and cold rolled)	Soluble oil Mineral lard oil Sulphurized oil	Mineral lard oil Sulphurized oil Soluble oil	Soluble oil Mineral lard oil
Tool steel (carbon and high-speed)	Soluble oil Mineral lard oil Sulphurized oil	Lard oil	Sulphurized oil Lard oil
Alloy steel	Soluble oil Mineral lard oil Sulphurized oil	Soluble oil Sulphurized oil Mineral lard oil	Sulphurized oil Lard oil
Brass and bronze	Dry Lard oil Kerosene mixture	Dry Soluble oil Lard oil	Soluble oil Lard oil
Copper	Dry Mineral lard oil Kerosene Soluble oil	Soluble oil Lard oil	Soluble oil Lard oil
Aluminum	Soluble oil Kerosene Lard oil	Soluble oil Kerosene Mineral oil	Soluble oil Kerosene and lard oil
Monel metal	Lard oil Soluble oil	Lard oil Soluble oil	Lard oil
Malleable iron	Dry Soda water	Dry Soda water	Lard oil Soda water
Cast iron	Dry Air jet Soluble oil	Dry Soluble oil Mineral lard oil	Dry Sulphurized oil Mineral lard oil

Chemical cutting fluids can be used for most of the above operations. Follow the manufacturer's recommendations for use and mixture.

CUTTING FLUIDS

The main purpose of cutting fluids is to reduce the amount of heat generated during the drilling operation, which dulls the drill quickly and produces a poor finish in the hole. Cutting fluids keep the cutting edges cool, prolong the life of the cutting tool, permit the use of faster cutting speeds, aid in the chip removal, and produce a better surface finish in the hole.

TYPES OF CUTTING FLUIDS

There are many types of cutting fluids available (Table 10-3), but they fall into three general categories: cutting oils, emulsifiable oils, and chemical cutting fluids.

Cutting oils are generally mineral oils, which contain certain additives to improve the drill's cutting action (Fig. 10-12). Cutting oils fall into two general categories: *active cutting oils*, used for drilling ferrous metals; and *inactive cutting oils*, used for drilling nonferrous metals.

Emulsifiable or soluble oils are oils which are first mixed with an emulsifier, which breaks up the oil into minute particles. This concentrate is then mixed with water at a ratio as high as fifty parts of water to one part of concentrate. Because of the good cooling qualities of soluble oils, they are suitable for high cutting speeds and where considerable heat is generated.

Chemical cutting fluids contain water-soluble chemicals which give the fluid cooling, lubricating, and anti-weld properties. These synthetic cutting fluids are efficient and cleaner than cutting oils and soluble oils, and they are finding wide use.

DRILL PRESS SAFETY

The drill press is probably the most common machine tool used in industry, school shops, and the home. Because it is so common, good safety practices which can prevent accidents are too often ignored. Before operating a drill press, the operator should become familiar with the safety rules in order to avoid an accident and personal injury.

Safety Rules

1. Never wear any loose clothing or ties around a machine.
   NOTE: Roll up the sleeves to above the elbow to prevent getting them caught in the machine (Fig. 10-13).

Fig. 10-13
Loose clothing can easily be caught in the revolving cutting tools or moving parts of machinery.

Fig. 10-12
Cutting oils are generally mineral oils with certain additives.

2. Long hair should be protected by a hair net or shop cap to prevent it from becoming caught in the revolving parts on a drill press.
3. Never wear rings, watches, or bracelets while working in a machine shop (Fig. 10-14).

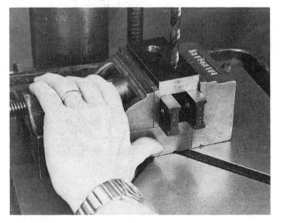

Fig. 10-14
Wearing rings and watches can be dangerous in a machine shop.

4. Always wear safety glasses when operating any machine.
5. Never attempt to set the speeds, adjust or measure the work until the machine is completely stopped.
6. Keep the work area and floor clean and free of oil and grease (Fig. 10-15).

Fig. 10-15
Poor housekeeping can lead to accidents.

7. *Never* leave a chuck key in a drill chuck *at any time*.
8. Always use a brush to remove chips.
9. Never attempt to hold work by hand when drilling holes larger than 12 mm (1/2 in.) in diameter. Use a clamp or table stop to prevent the work from spinning.
10. Ease up on the drilling pressure as the drill breaks through the workpiece. This will prevent the drill from pulling into the work and breaking.
11. Always remove the burrs from a hole which has been drilled.

DRILLING CENTRE HOLES

Work that is to be turned between the centres on a lathe must have a centre hole drilled in each end so that the work may be supported by the lathe centres. Although centre holes in the work are more easily and accurately drilled on a lathe, they are often machined on a drill press because of the shape of the workpiece or the equipment available.

After the centre locations have been laid out on both ends of the workpiece, a *combined drill and countersink*, commonly called a *centre drill*, is used to drill the centre holes.

Two types of centre drills are available: the plain or regular type (Fig. 10-16A) and the bell type (Fig. 10-16B). The

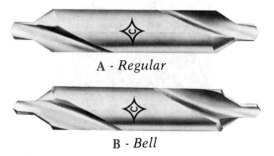

A - *Regular*

B - *Bell*

Courtesy Cleveland Twist Drill (Canada) Ltd.

Fig. 10-16
Two types of centre drills.

regular type has a 60° angle with a small drill located on the end. The bell type is similar but has a secondary bevel near the large diameter, which produces a clearance angle near the top of the hole. This clearance angle prevents the top edge of the 60° bearing surface from becoming burred or damaged. Centre drills are available in a wide variety of sizes to suit different diameters of work. Table 10-4 lists information regarding regular and bell-type centre drills and the diameters for which each should be used.

In order to provide a good bearing surface for the work on the lathe centres, it is important that the centre holes be drilled to the correct size (see Table 10-4). The centre hole illustrated in Fig. 10-17A is too shallow and will not provide an adequate bearing surface. The centre hole in Fig. 10-17B is too deep and will not allow the taper of the lathe centre to contact the taper of the centre hole. The centre hole illustrated in Fig. 10-17C is drilled to the proper depth, which provides a good bearing surface for the lathe centres.

To Drill a Lathe Centre Hole

Centre holes should be as smooth and accurate as possible, to provide a good bearing surface and reduce the friction and wear between the work and lathe centres.

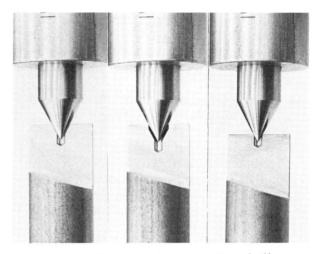

Properly drilled   Too deep   Too shallow

Courtesy Kostel Enterprises Ltd.

Fig. 10-17
Improperly and properly drilled centre holes.

1. Check the centre hole layout to be sure that it is correct.
2. Obtain the correct size centre drill to suit the diameter of the work being drilled (Table 10-4).

TABLE 10-4 CENTRE DRILL SIZES						
Size		Work Diameter		Diameter of Countersink C	Drill Point Diameter	Body Size
Regular Type	Bell Type	Millimetres	Inches			
1	11	3—8	3/16—5/16	3/32	3/64	1/8
2	12	9.5—12.5	3/8—1/2	9/64	5/64	3/16
3	13	15—20	5/8—3/4	3/16	7/64	1/4
4	14	25—40	1—1-1/2	15/64	1/8	5/16
5	15	50—75	2—3	21/64	3/16	7/16
6	16	75—100	3—4	3/8	7/32	1/2
7	17	100—125	4—5	15/32	1/4	5/8
8	18	150 and over	6 and over	9/16	5/16	3/4

3. Fasten the centre drill in the drill chuck. To prevent breakage, do not have more than 12 mm (1/2 in.) of the centre drill extending beyond the drill chuck.
4. Set the drill press to the proper speed for the size of the centre drill being used.
   A speed of 1200 to 1500 *r/min* is suitable for most centre drills.
5. Fasten a clamp or table stop on the left side of the table to prevent the vise from swinging during the drilling operation (Fig. 10-18).
6. Set the vise on its side on a clean drill press table.
7. Press the work firmly against the bottom of the vise, and then tighten it securely.
8. With the vise against the table clamp or stop, locate the centre punch mark of the work under the drill point.
9. Start the drill press and with the hand feed lever carefully feed the centre drill into the work.
   CAUTION: *Too fast a feed pressure may break the drill point.*
10. Frequently raise the drill from the work and apply a few drops of cutting fluid.
11. Continue drilling until the top of the countersunk hole (C) is the correct size (Table 10-4).
12. If the centre hole is not too smooth, apply a little cutting fluid and lightly bring the centre drill into the hole again.

To Spot a Hole with a Centre Drill

The chisel edge at the end of the web of a drill is generally larger than the centre punch mark on the work, and therefore it is difficult to start a drill at the exact location. To prevent a drill from wandering off centre, it is considered good practice to first spot every centre punch mark with a centre drill. The small point on the centre drill will accurately follow the centre punch mark and provide a guide for the larger drill which will be used.

1. Mount a small size centre drill in the drill chuck.
2. Mount the work in a vise or set it on the drill press table. *Do not clamp the work or the vise.*
3. Set the drill press speed to about 1500 r/min.
4. Bring the point of the centre drill into the centre punch mark, and allow the work to centre itself with the drill point (Fig. 10-19).
5. Continue drilling until about one-third of the tapered section of the centre drill has entered the work.
6. Spot all the holes which are to be drilled.

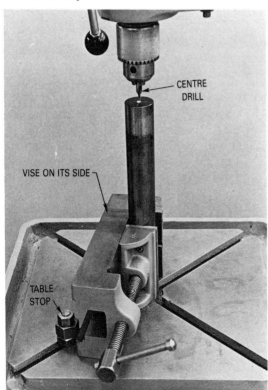

Courtesy Kostel Enterprises Ltd.

Fig. 10-18
The work is held in a vise set on its side for drilling centre holes.

Courtesy Kostel Enterprises Ltd.

Fig. 10-19
A centre drill is used to spot the location of a hole accurately.

DRILLING A HOLE

To drill a hole accurately and safely, it is wise to always observe the following points:

- Measure the drill with a micrometer or gauge to be sure that it is the correct size, especially if the hole is to be reamed or tapped later.
- Clamp the work properly to prevent inaccurate work and accidents.
- Always use a sharp drill which is correctly ground for the material being drilled.
- Set the drill press to the proper speed and feed to avoid damaging the twist drill, the machine, or having an accident.

To Drill a Hole in Work Held in a Vise

The most common method of holding small workpieces is by means of a vise, which may be held by hand against a table stop or clamped to the table. When drilling holes larger than 12 mm (1/2 in.) in diameter, the vise should be clamped to the table.

1. Spot the hole location with a centre drill.
2. Mount the correct size drill in the drill chuck.
3. Set the drill press to the proper speed for the size of drill and the type of material to be drilled.
4. Fasten a clamp or stop on the left side of the table (Fig. 10-20).

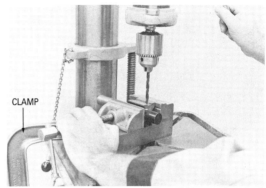

Courtesy Kostel Enterprises Ltd.

Fig. 10-20
The vise can be held against a table stop by hand when drilling holes up to about 12 mm (1/2 in.) in diameter.

5. Mount the work on parallels in a drill vise, and tighten it securely.
6. With the vise against the table stop, locate the spotted hole under the centre of the drill.
7. Start the drill press spindle and begin to drill the hole.
   (a) For holes up to 12 mm (1/2 in.) in diameter, hold the vise against the table stop by hand (Fig. 10-20).
   OR
   (b) For holes over 12 mm (1/2 in.) in diameter
      (i) Lightly clamp the vise to the table with another clamp (Fig. 10-21).
      (ii) Drill until the full drill point is into the work.
      (iii) With the drill revolving, keep the drill point in the work and tighten the clamp, holding the vise securely.
8. Raise the drill occasionally and apply cutting fluid during the drilling operation.

DRILLING A HOLE 111

9. Ease up on the drilling pressure as the drill starts to break through the workpiece.

Courtesy Kostel Enterprises Ltd.

Fig. 10-21
The vise should be clamped to the table when drilling holes over about 12 mm (1/2 in.) in diameter.

To Drill Work Fastened to a Drill Table

Sometimes, because of the size or shape of the workpiece, it is necessary to clamp the work directly to the table. When work is clamped to the table, always be sure to use parallels between the table and work, so that the drill will not cut into the table.

1. Spot the hole location with a centre drill.
2. Mount the correct drill size and set the drill press speed and feed.
3. Set the work on a suitable set of parallels.
4. Locate the spotted hole under the point of the twist drill.
5. Select suitable clamps, step blocks, and bolts, and position them on the work as shown in Fig. 10-22.
6. Lightly tighten each clamp.
7. Start the drill revolving and feed the drill until about one-half the drill point has entered the work.

If the work is lightly clamped, the spotted hole will align itself with the drill point.

8. While the revolving drill is still in contact with the work, *tighten both clamps securely.*
9. Occasionally apply cutting fluid during the drilling operation.
10. Ease up on the drilling pressure as the drill begins to break through the work.

Courtesy Kostel Enterprises Ltd.

Fig. 10-22
The clamps are correctly set for clamping a workpiece to the drill press table.

To Drill a Pilot Hole for Large Drills

As the size of a drill increases, the thickness of the web of the drill increases to strengthen the drill. The thicker web results in a longer chisel point on the drill. Since the chisel point does not cut, larger drills require more pressure to feed them into the work. To relieve some of this drilling pressure and provide a guide for the larger drill to follow, a pilot hole is first drilled in the work at the location of the hole.

The size of pilot hole drilled should only be slightly larger than the thickness of the web of the drill to be used (Fig. 10-23). If the pilot hole is drilled too large, the following drill may chatter, drill the hole out of round, or damage the top of the

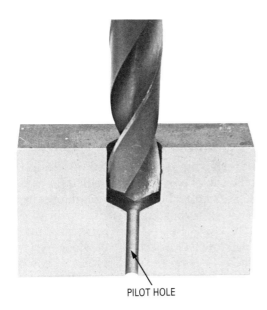

PILOT HOLE

Courtesy Kostel Enterprises Ltd.

Fig. 10-23
A pilot hole reduces drilling pressure and prevents a larger drill from wandering.

hole. Care must be used to drill the pilot hole on centre, because the larger drill *will follow the pilot hole*. This method may also be used to drill average-size holes when the drill press is small and does not have sufficient power to drive the drill through the solid metal.

To Drill Round Work in a V-Block

V-blocks are used to hold round work accurately for drilling. The round material is seated in the accurately machined groove, and small diameters are held securely with a U-shaped clamp. Larger work may be supported by a V-block in a drill vise, or it may be mounted on V-blocks and clamped to the table.

1. Select a V-block to suit the diameter of round work to be drilled. If the work is long, use a pair of V-blocks.
2. Mount the work in the V-block, and then rotate it until the centre punch mark is in the centre of the workpiece.

Courtesy Kostel Enterprises Ltd.

Fig. 10-24
Using a square and rule to align the centre punch mark on round work.

Check that the distance from both sides is equal with a rule and square (Fig. 10-24).
3. Tighten the U-clamp securely on the work in the V-block.

OR

Hold the work and V-block in a vise as shown in Fig. 10-25.
4. Spot the hole location with a centre drill.
5. Mount the proper size drill, and set the machine to the correct speed.
6. Drill the hole, being sure that the drill does not hit the V-block or vise when it breaks through the work.

Courtesy Kostel Enterprises Ltd.

Fig. 10-25
The work and V-block held in a vise and ready for drilling.

DRILLING A HOLE 113

DRILL PRESS OPERATIONS

With the use of various cutting tools, a variety of operations can be performed on a drill press. Some of the more common drill press operations are briefly described below.

COUNTERSINKING

Countersinking is the process of enlarging the top of a hole to the shape of a cone. Countersinks (Fig. 10-26A) are available with included angles of 60° and 82°. The 60° countersink is used for producing lathe centre holes, while the 82° countersink is used to produce the tapered hole which accommodates a flat head bolt or machine screw (Fig. 10-26B). Countersinks may also be used to remove burrs from the top of a drilled hole.

When countersinking for a flat head machine screw, the hole should be countersunk so that the head of the screw is flush with the top of the work surface (Fig. 10-26B). The speed for countersinking is generally about one-quarter of the recommended drilling speed.

COUNTERBORING

Counterboring is the operation of enlarging the top of a previously drilled hole (Fig. 10-27). Counterbores are available in a variety of types and sizes, with pilots which may be either solid or interchangeable. Holes are counterbored to create an enlarged hole with a square shoulder to accommodate the head of a bolt or cap screw, or the shoulder on a pin. The speed for counterboring is usually about one-quarter of the drilling speed.

Courtesy Cleveland Twist Drill (Canada) Ltd.

Fig. 10-26A
A countersink.

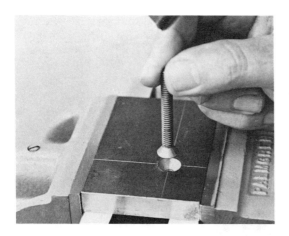

Courtesy Kostel Enterprises Ltd.

Fig. 10-26B
Work properly countersunk to suit the head of a flat-head machine screw.

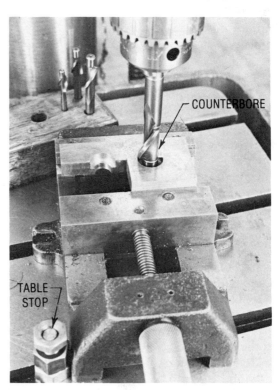

Courtesy Kostel Enterprises Ltd.

Fig. 10-27
Counterboring enlarges the top of a hole.

Courtesy Kostel Enterprises Ltd.

Fig. 10-28
Reaming brings a hole to size and produces a smooth hole.

REAMING

The purpose of reaming is to bring a drilled or bored hole to size and shape, and to produce a good surface finish in the hole (Fig. 10-28). Speed, feed, and reaming allowance are three factors that can affect the accuracy of a reamed hole. Approximately 0.4 mm (1/64 in.) is left for reaming holes up to 12.5 mm (1/2 in.) in diameter; 0.8 mm (1/32 in.) is recommended for holes over 12.5 mm (1/2 in.) diameter. The speed for reaming is generally about one-half of the drilling speed.

There are two types of reamers used in machine shop work: hand reamers and machine reamers. Hand reamers have a square on one end and are used to remove no more than 0.12 mm (0.005 in.) from a hole. Machine reamers have a straight or tapered shank and are used under power.

BORING

Boring is the process of enlarging a previously drilled or cored hole to produce a straight hole and bring it to an accurate size. Most of the boring in a drill press is performed with a single-point cutting tool mounted in a boring bar, which is held in the drill press spindle (Fig. 10-29). The speed for boring is the same as that used for drilling a hole of the same size.

A drill press should only be used for boring if it is impractical to use another machine for this operation. In order to bore effectively in a drill press, the table *must be* equipped with a hole in the centre, which serves as a guide or support for the end of the boring bar.

Fig. 10-29
Boring produces a true round hole that is accurate to size.

SPOT-FACING

Spot-facing is the operation of smoothing and squaring the surface around the top of a hole to provide a flat seat for the head of a cap screw or nut. A boring bar, with a pilot on its end to fit the hole, is fitted with a double-edged cutting tool (Fig. 10-30). When spot-facing, it is important that the work be securely clamped and the drill press set to about one-quarter of the drilling speed.

Courtesy Kostel Enterprises Ltd.

Fig. 10-30
Spot-facing produces a flat seat around the top of a hole.

TAPPING

Tapping in a drill press can be performed either by hand or with the use of a tapping attachment. The advantage of tapping a hole in a drill press is that the tap can be started squarely and kept that way throughout the entire length of the hole being threaded.

1. Mount the work on suitable parallels, and lightly clamp the work to the drill press table. Work may be held in a vise which is lightly clamped to the table.
2. Mount a centre drill in the drill chuck, and adjust the drill press table or the work until the centre punch mark aligns with the point of the centre drill.
3. Spot the hole with a centre drill and then tighten the clamps securely.
4. Drill the hole to the correct tap drill size for the size of tap to be used.

    The work or table must not be moved after drilling the hole; otherwise the alignment will be disturbed and the tap will not enter squarely.
5. Mount a stub centre in the drill chuck (Fig. 10-31).

    OR

    Remove the drill chuck and mount a special centre in the drill press spindle.
6. Fasten a tap wrench on the correct size tap and place it into the hole.
7. Lower the drill press spindle until the stub centre point fits into the centre hole in the end of the tap shank.
8. Turn the tap wrench to start the tap into the hole, and at the same time keep the centre in light contact with the tap.
9. Continue to tap the hole in the usual manner while keeping the tap aligned by applying light pressure on the drill press downfeed lever.

A tapping attachment may be mounted in the drill press spindle to rotate the tap by power. Special two- or three-fluted gun taps are used for tapping under power because of their ability to clear chips. The speed for tapping under power generally ranges from 60 to 100 *r/min*.

TEST YOUR KNOWLEDGE

1. Name six operations which can be performed on a drill press.

Drill Press Types and Construction

2. Name two common types of drill presses.
3. List four main parts of a drill press and give one use for each.
4. State the purpose of the following parts:
   (a) spindle   (c) hand feed lever
   (b) spindle sleeve (d) depth stop

Tool-Holding Devices

5. For what purpose are drill chucks used?
6. How do the jaws on most drill chucks operate?
7. State the purpose of:
   (a) a drill sleeve   (b) a drill socket
8. How should a taper shank tool be mounted in a drill press spindle?
9. How should a taper shank tool be removed from a drill press spindle?

Twist Drills

10. Describe the most common twist drill manufactured.
11. Of what two materials are twist drills manufactured?

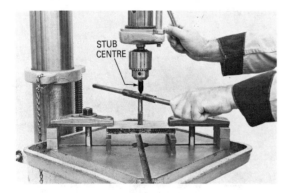

Courtesy Kostel Enterprises Ltd.

Fig. 10-31
A stub centre being used to keep a tap straight in a drill press.

12. State the purpose of each part:
    (a) tang   (d) body clearance
    (b) flutes   (e) web
    (c) margin   (f) lip clearance
13. What is the recommended point angle and lip clearance for general purpose drilling?

Systems of Drill Sizes
14. Name the three systems used to designate inch drill sizes. In what ranges are metric drills available?
15. What size are the following drills: #1, #29, #60, Letters A, F, K, Z?
16. Name two methods of measuring the size of a drill.

Cutting Speeds and Feeds
17. Why are speeds and feeds so important to the life of a cutting tool and the production rate?
18. Define: (a) cutting speed
            (b) metres per minute
            (c) *r/min*
19. Calculate the *r/min* required to drill machine steel (100 CS) with the following drills: 6 mm, 5/8 in., 35 mm, 2 in.
20. Name two factors which may affect the calculated drill speed.
21. Define drilling speed.
22. Explain the effects of: (a) too coarse, and (b) too fine, a drilling feed.

Cutting fluids
23. List four purposes of a cutting fluid.
24. State where the following should be used:
    (a) cutting oils
    (b) emulsifiable or soluble oils
    (c) chemical cutting fluids

Drill Press Safety
25. Why is loose clothing and long hair dangerous around a machine?
26. Name three other safety rules you consider most important.

Drilling Centre Holes
27. State the difference between a regular type and a bell type centre drill.
28. Why should centre holes not be drilled
    (a) too shallow?   (b) too deep?
29. Why is it important that centre holes be drilled as smoothly and accurately as possible?
30. How much of the centre drill should extend beyond the drill chuck?
31. At what speed should the drill press be set for centre drilling?
32. Why is a centre drill suitable for spotting a hole?
33. How deep should holes be spotted?

Drilling a Hole
34. List three important points which should be observed in order to drill a hole accurately and safely.
35. When should a vise be clamped to the table?
36. What is the purpose of fastening a stop to the left side of the table?
37. When should the drilling pressure be eased?
38. Why should parallels be used between the table and the workpiece?
39. Where should the bolts and step blocks be placed when clamping work to the table?
40. What size pilot hole should be drilled in relation to the larger drill which will be used?

To Drill Work in a V-block
41. Explain the procedure for locating the centre punch mark in the centre of a V-block.
42. How can work be held in a V-block for drilling a hole?

Drill Press Operations
43. What is the purpose of countersinking?
44. What countersink should be used to provide a seat for a flat head machine screw?
45. State two purposes of reaming.
46. How may hand and machine reamers be identified?
47. When tapping in a drill press, why should the work or table not be moved after the hole is drilled?
48. How is the tap kept in alignment during the tapping process?

CHAPTER 11
THE ENGINE LATHE

Courtesy Clausing Division of Atlas Press Company

The lathe, probably one of the earliest machine tools, is one of the most versatile and widely used machines. Because a large percentage of the metal cut in a machine shop is cylindrical, the basic lathe has led to the development of turret lathes, screw machines, boring mills, numerically controlled lathes, and turning centres. The progress in the design of the basic engine lathe and its related machines has been responsible for the development and production of thousands of products we use and enjoy every day.

The main function of the engine lathe is to turn cylindrical shapes and workpieces (Fig. 11-1). This is done by rotating the metal held in a work-holding device while a cutting tool is forced against its circumference. Figure 11-2 shows the cutting action of a tool on work being machined in a lathe. Some of the common operations performed on a lathe are: facing, taper turning, parallel turning, thread cutting, knurling, boring, drilling, and reaming. The engine lathe is the backbone of a machine shop, and a thorough knowledge of it is essential for any machinist.

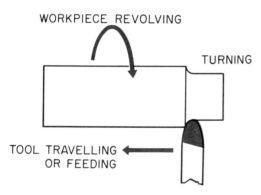

Fig. 11-1
The cutting action of an engine lathe.

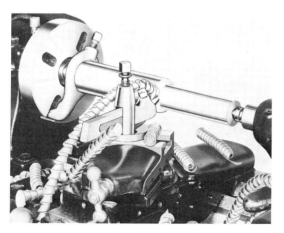

Courtesy South Bend Lathe, Inc.

Fig. 11-2
A roughing cut on a workpiece held between centres.

SIZE OF THE ENGINE LATHE

The size of an engine lathe is determined by the maximum diameter of work which may be revolved or swung over the bed (Fig. 11-3).

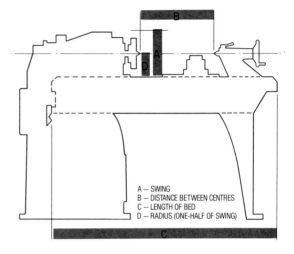

Fig. 11-3
The size of a lathe is determined by the swing and the length of the bed.

Lathes found in school shops may have a swing of 230 to 330 mm (9 to 13 in.) and a bed length from 500 to 1500 mm (20 to 60 in.). Lathes used in industry may have swings of 230 to 760 mm (9 to 30 in.) and a capacity of 400 mm to 3-1/2 m (16 in. to 12 ft.). A typical lathe may have a 330 mm (13 in.) swing, a 2 m (6 ft.) long bed, and a capacity to turn work 1 m (36 in.) between centres.

LATHE PARTS

The main function of a lathe is to provide a means of rotating a workpiece against a cutting tool, thereby removing metal. All lathes, regardless of design or size, are

basically the same (Fig. 11-4), and serve three functions. They provide:
1. A support for the lathe accessories or the workpiece.
2. A way of holding and revolving the workpiece.
3. A means of holding and moving the cutting tool.

BED

The bed is a heavy rugged casting made to support the working parts of the lathe. On its top section are machined ways that guide and align the major parts of the lathe. Many lathes are made with flame-hardened and ground ways to reduce wear and maintain accuracy.

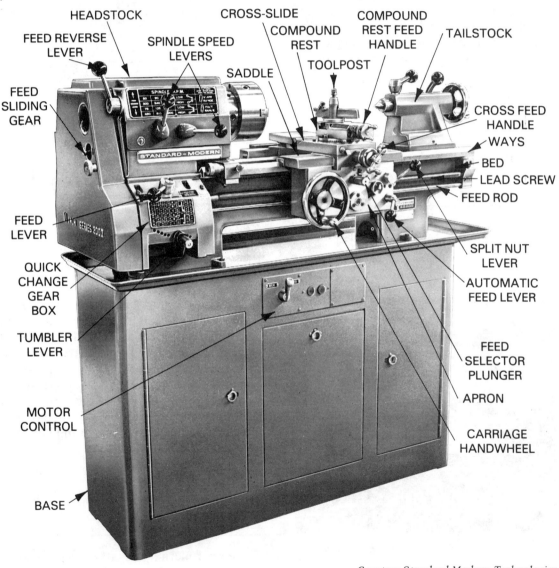

Courtesy Standard-Modern Technologies

Fig. 11-4
The main parts of an engine lathe.

120 CHAPTER 11 / THE ENGINE LATHE

HEADSTOCK

The headstock is clamped on the left-side of the bed. The *headstock spindle*, a hollow cylindrical shaft supported by bearings, provides a drive from the motor to work-holding devices. A live centre and sleeve, a face plate, or a chuck can be fitted to the spindle nose to hold and drive the work. The live centre has a 60° point which provides a bearing surface for the work to turn between centres.

Most modern lathes are geared-head, and the spindle is driven by a series of gears in the headstock. This arrangement permits several spindle speeds to accommodate different types and sizes of work.

The *feed reverse lever* can be placed in three positions. The top position provides a forward direction to the *feed rod* and *lead screw*, the centre position is neutral, and the bottom position reverses the direction of the feed rod and lead screw.

QUICK-CHANGE GEARBOX

The *quick-change gearbox*, which contains a number of different-sized gears, provides the feed rod and lead screw with various speeds for turning and thread-cutting operations. The feed rod and lead screw provide a drive for the carriage when either the *automatic feed lever* or the *split-nut lever* is engaged.

CARRIAGE

The carriage supports the cutting tool and is used to move it along the bed of the lathe for turning operations. The carriage consists of three main parts, the *saddle*, the *apron*, and the *cross-slide* (Fig. 11-5).

The *saddle*, an H-shaped casting mounted on the top of the lathe ways, supports the *cross-slide*, which provides a cross movement for the cutting tool. The *compound rest* is used to support the cutting tool and can be swivelled at any angle

Fig. 11-5
The main parts of a lathe carriage.

Courtesy Standard-Modern Technologies

for taper-turning operations. The cross-slide and the compound rest are moved by feed screws. Each of these is provided with a graduated collar to make accurate settings possible for the cutting tools.

The *apron* is fastened to the saddle and houses the feeding mechanisms, which provide an automatic feed to the carriage. The automatic feed lever is used to engage feeds to the carriage. The *apron handwheel* can be turned by hand to move the carriage along to the bed of the lathe. This handwheel is connected to a gear which meshes in a rack fastened to the bed of the lathe. The *feed directional plunger* can be shifted into three positions. The *in position* engages the longitudinal feed for the carriage. The *centre or neutral position* is used in thread-cutting to allow the split-nut lever to be engaged. The *out position* is used when an automatic crossfeed is required.

TAILSTOCK

The tailstock (Fig. 11-6) is made up of two units. The top half can be adjusted on the base by two adjusting screws for aligning the tailstock and headstock centres for parallel turning. These screws can also be used for offsetting the tailstock for taper-turning between centres. The tailstock can be locked in any position along the bed of the lathe by tightening the *clamp lever or nut*. One end of the dead centre is tapered to fit into the *tailstock spindle*, while the other end has a 60° point to provide a bearing support for work turned between centres. Other standard tapered tools, such as reamers and drills, can be held in the tailstock spindle. A *spindle binding lever* or lock handle is used to hold the tailstock spindle in a fixed position. The *tailstock handwheel* moves the tailstock spindle in or out of the tailstock casting. It can also be used to provide a hand feed for drilling and reaming operations.

SAFETY PRECAUTIONS

A lathe can be very dangerous if not operated properly, even though it is equipped with various safety guards and features. It is the duty of the operator to observe various safety precautions and prevent accidents. Everyone must realize that a clean and orderly area around a machine will go a long way in preventing accidents.

The following are some of the more important safety regulations which should be observed when operating a lathe:

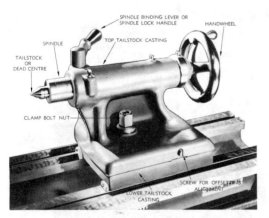

Courtesy Standard-Modern Technologies

Fig. 11-6
The parts of the tailstock assembly.

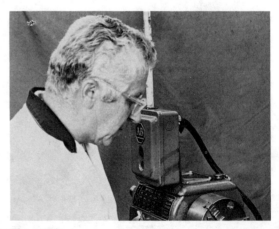

Fig. 11-7
Safety glasses must always be worn in a machine shop.

1. Always wear safety glasses when operating any machine (Fig. 11-7).
2. Never attempt to run a lathe until you are familiar with its operation.
3. Never wear loose clothing, rings, or watches when operating a lathe.
   NOTE: *These could be caught by the revolving parts of a lathe and cause a serious accident.*
4. Always stop the lathe before making measurements of any kind.
5. Always use a brush to remove chips.
   NOTE: *DO NOT handle them by hand; they are very sharp* (Fig. 11-8).

Fig. 11-9
Leaving a chuck wrench in a chuck may result in a serious injury.

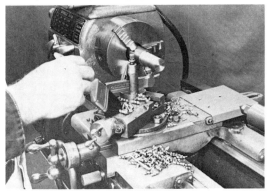

Fig. 11-8
Removing chips with your hand is a dangerous practice. Always use a brush.

6. Before mounting or removing accessories, always shut off the power supply to the motor.
7. Do not take heavy cuts on long slender pieces.
   CAUTION: *This could cause the work to bend and fly out of the machine.*
8. Do not lean on the machine. Stand erect, keeping your face and eyes away from flying chips.
9. Keep the floor around a machine clean and free of grease, oil, and other materials that could cause dangerous falls.
10. Never leave a chuck wrench in a chuck.
    CAUTION: *If the machine is started, the wrench will fly out and possibly injure someone* (Fig. 11-9).

CUTTING SPEED AND FEEDS

The cutting speed for lathe work may be defined as the rate at which a point on the circumference of the work passes the cutting tool. Cutting speed is expressed in metres per minute (m/min) or in feet per minute (ft./min). For example, if tool steel has a cutting speed of 18 m/min (60 ft./min) the lathe speed (r/min) should be set so that 18 m (60 ft.) of the work circumference passes the cutting tool in one minute. Too slow a lathe speed will result in the loss of valuable time, while too fast a speed will cause the cutting tool to break down quickly, resulting in time being wasted resharpening the cutting tool. The recommended cutting speed (*CS*) for various material is listed Table 11-1. These cutting speeds have been determined by cutting tool and metal manufacturers as being the best for production rates and cutting tool life.

SPINDLE SPEED CALCULATIONS

To be able to calculate the number of revolutions per minute at which to set a lathe, the diameter of the work and the cutting speed of the material must be known. The revolutions per minute at which the lathe should be set can be found

TABLE 11-1 LATHE CUTTING SPEEDS IN METRES PER MINUTE AND FEET PER MINUTE USING A HIGH-SPEED TOOLBIT

	Turning and Boring				Threading	
	Rough Cut		Finish Cut			
Material	m/min	ft./min	m/min	ft./min	m/min	ft./min
Machine Steel	27	90	30	100	11	35
Tool Steel	21	70	27	90	9	30
Cast Iron	18	60	24	80	8	25
Bronze	27	90	30	100	8	25
Aluminum	61	200	93	300	18	60

by applying one of the following simplified formulas:

The standardized spindle speed formulae (and their development) may be found in Chapter 10, *Drill Presses*.

Inch Calculations

The spindle speed of a lathe, where the workpiece dimensions are given in inches, is:

$$r/min = \frac{CS \times 4}{D}$$

CS = cutting speed of the metal in ft./min.
D = diameter of the workpiece in inches

EXAMPLE: Calculate the *r/min* required to finish turn a 2 in. diameter piece of machine steel. Table 11-1 lists the cutting speed for machine steel as 100.

$$r/min = \frac{CS \times 4}{D}$$
$$= \frac{100 \times 4}{2}$$
$$= 200$$

Metric Calculations

The simplified formula for determining the spindle speed when the cutting speed is given in metres is:

$$r/min = \frac{CS \times 320}{D}$$

CS = cutting speed in m/min
D = diameter of work in millimetres

EXAMPLE: Calculate the *r/min* required to finish turn a 60 mm diameter piece of machine steel.

$$r/min = \frac{30 \times 320}{60}$$
$$= 160$$

Since most lathes are provided with a limited number of speed settings, this simplified formula is acceptable for all calculations. When it is not possible to set the lathe to the exact spindle speed, always set it to the next lower speed.

SETTING LATHE SPEEDS

Engine lathes are designed to operate at various spindle speeds, for machining different-sized diameters and types of material. These speeds are measured in revolutions per minute and are changed by means of gear levers or a variable speed adjustment. When setting the spindle speed, always set the machine as close as possible to the calculated speed, but never higher. If the cutting action is satisfactory, the speed may be increased slightly; however, if the cutting action is not

Courtesy Standard-Modern Technologies

Fig. 11-10
The speed of the geared-head lathe is regulated by the position of the headstock levers.

satisfactory or the work vibrates or chatters, reduce the speed and increase the feed.

On the geared-head lathe (Fig. 11-10) speeds are changed by moving the speed levers into their proper positions according to the *r/min* chart that is fastened to the headstock. While shifting the lever positions, place one hand on the drive plate or chuck, and turn it slowly by hand. This will enable the levers to engage the gear teeth without clashing.

NOTE: Never change speeds when the lathe is running.

Some lathes are equipped with a variable speed headstock, and any speed within the range of the machine can be set. The spindle speed is set while the lathe is running by turning a speed control knob until the desired speed is indicated on the speed dial.

LATHE FEED

The feed of a lathe is defined as the distance the cutting tool advances along the length of the work for every revolution of the spindle. For example, if the lathe is set for a 0.20 mm (0.008 in.) feed, the cutting tool will travel along the length of the work 0.20 mm (0.008 in.) for every complete turn that the work makes. The feed of an engine lathe is dependent upon the speed of the lead screw or feed rod. This is controlled by the change gears in the *quick-change gearbox* (Fig. 11-11A).

Whenever possible, only two cuts should be taken to bring a diameter to size: a roughing cut and a finishing cut. Since the purpose of a roughing cut is to remove excess material quickly and surface finish is not too important, a coarse feed should be used. The finishing cut is used to bring the diameter to size and produce a good surface finish; therefore a fine feed should be used. For general purpose machining, a 0.25 to 0.40 mm (or a 0.010 to 0.015 in.) feed for roughing and a 0.07 to 0.12 mm (or a 0.003 in. to 0.005 in.) feed for finishing is recommended. Table 11-2 lists the recommended feeds for cutting various materials when using a high-speed steel cutting tool.

	TABLE 11-2 FEEDS FOR VARIOUS MATERIALS (USING A HIGH-SPEED CUTTING TOOL)			
	Rough Cuts		Finish Cuts	
Material	Millimetres	Inches	Millimetres	Inches
Machine Steel	0.25—0.50	0.010—0.020	0.07—0.25	0.003—0.010
Tool Steel	0.25—0.50	0.010—0.020	0.07—0.25	0.003—0.010
Cast Iron	0.40—0.65	0.015—0.025	0.13—0.30	0.005—0.012
Bronze	0.40—0.65	0.015—0.025	0.07—0.25	0.003—0.010
Aluminum	0.40—0.75	0.015—0.030	0.13—0.25	0.005—0.010

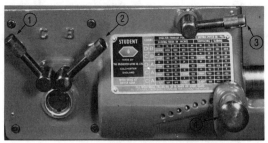

Courtesy Colchester Lathe Co.

Fig. 11-11A
The quick-change gear box controls the feed of the lathe.

To Set the Lathe Feed

1. From the quick-change gearbox chart (Fig. 11-11B), select the amount of feed required.
2. Move the tumbler lever #4 (Fig. 11-11A) into the hole directly below the row in which the selected amount of feed is found.
3. Follow the row in which the selected feed is found to the left, and set the *feed change levers* (#1 and #2) to the letters indicated on the feed chart.
4. Set lever #3 (lead screw engaging lever) to the down position.

   NOTE: Before turning on the lathe, be sure all levers are fully engaged by turning the headstock spindle by hand. If all the levers are correctly engaged, the feed rod should turn as the spindle is revolved.

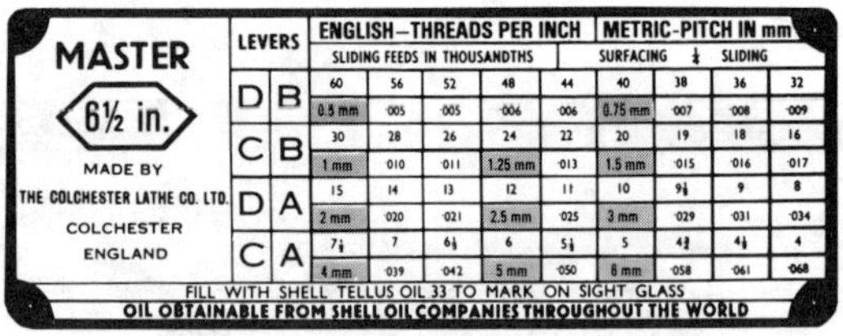

MASTER 6½ in. MADE BY THE COLCHESTER LATHE CO. LTD. COLCHESTER ENGLAND	LEVERS		ENGLISH—THREADS PER INCH					METRIC-PITCH IN mm			
			SLIDING FEEDS IN THOUSANDTHS					SURFACING ½ SLIDING			
	D	B	60	56	52	48	44	40	38	36	32
			0.5 mm	·005	·005	·006	·006	0.75 mm	·007	·008	·009
	C	B	30	28	26	24	22	20	19	18	16
			1 mm	·010	·011	1.25 mm	·013	1.5 mm	·015	·016	·017
	D	A	15	14	13	12	11	10	9½	9	8
			2 mm	·020	·021	2.5 mm	·025	3 mm	·029	·031	·034
	C	A	7½	7	6½	6	5½	5	4¾	4½	4
			4 mm	·039	·042	5 mm	·050	6 mm	·058	·061	·068

FILL WITH SHELL TELLUS OIL 33 TO MARK ON SIGHT GLASS
OIL OBTAINABLE FROM SHELL OIL COMPANIES THROUGHOUT THE WORLD

Feeds & threads available from Master gearbox

Courtesy Colchester Lathe Co.

Fig. 11-11B
The chart on the quick-change gear box indicates the feeds and threads that can be set on a lathe.

DEPTH OF CUT

The depth of cut in lathe work may be defined as the chip or cut that is taken by the cutting tool. Figure 11-12 shows a 0.020 in. cut being taken from a piece of work 1.000 in. in diameter, reducing the work by 0.040 in. to 0.960 in. in diameter. If a 1 mm depth of cut was taken from a 25 mm diameter, the work would be reduced to 23 mm. In rough turning, the depth of cut depends upon the condition of the machine, the type of cutting tool used, and the rigidity of the workpiece. In finish turning, the depth of cut, which should never be less than 0.10 mm (0.004 in.), is dependent on the type of work and finish required.

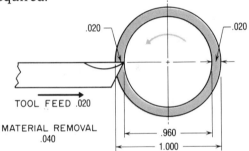

Fig. 11-12
Since the work in a lathe revolves, a 0.020 in. depth of cut reduces the diameter 0.040 in.

CUTTING TOOLS AND TOOLHOLDERS

To machine metal in a lathe, a cutting tool called a *toolbit* is used. Toolbits used in school shops are usually made of high-speed steel. Carbide toolbits are generally used in industry because they can withstand the heat and friction created by high cutting speeds. They are not used extensively in school shops because of their cost and the higher-horsepower equipment required to machine work with carbide cutting tools. Toolbits are made in a variety of sizes and shapes for use with different size machines and different applications.

In order to cut the various shapes desired, specially shaped toolbits are used (Fig. 11-13). Left-hand toolbits have their cutting edge on the right-hand side and are used for turning work toward the tailstock. Right-hand cutting tools have the cutting edge on the left-hand side and are used for cutting toward the headstock. A lathe cutting tool is generally known by the operation it performs. For example, a roughing tool is used to rough-turn work, a threading tool is used for thread cutting, etc.

To produce various surfaces, faces, and forms, the cutting edge of the toolbit must be precisely ground. The cutting edge must be shaped to the proper form and then relieved with clearance angles to allow the edge to cut into the metal. The *end-relief* (clearance) angle is ground on the end of the toolbit to allow the point to be fed into the work. The *side relief* (clearance) angle is ground below the cutting edge to allow the toolbit to be fed lengthwise along the work. See "Grinding a Lathe Toolbit" in chapter 14, *Grinders*.

TOOLHOLDERS AND TOOLPOSTS

A variety of lathe toolholders and toolposts are available to suit various operations, manufacturing sequences, or the types of cutting tools being used.

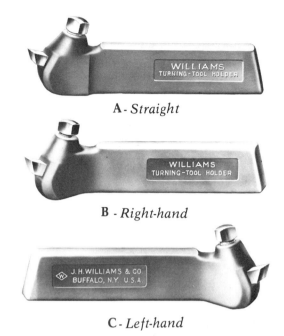

A - *Straight*

B - *Right-hand*

C - *Left-hand*

Fig. 11-14 *Courtesy J.H. Williams & Co.*
Common toolholders.

Toolbits can be used in three positions. For general turning, a straight toolholder (Fig. 11-14A) is used. When machining work to the right or squaring a right-hand surface, a right-hand toolholder (Fig. 11-14B) is used. When turning work toward the headstock or squaring a left-hand surface, a left-hand toolholder (Fig. 11-14C) is used.

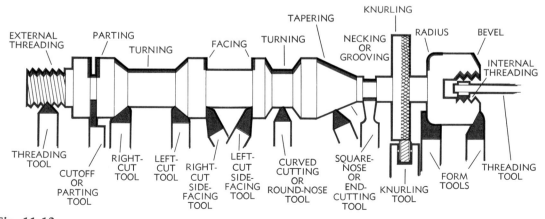

Fig. 11-13
Common cuts made by cutting tools.

CUTTING TOOLS AND TOOLHOLDERS 127

STANDARD (ROUND) TOOLPOST

The standard or round toolpost (Fig. 11-15) is generally supplied with the lathe. It consists of a round toolpost which fits into the T-slot of the compound rest. A concave ring, rocker, and toolpost screw provide a means of adjusting and holding the tool to the proper height.

Fig. 11-15
A standard toolpost.

TURRET-TYPE TOOLPOST

Turret-type toolposts (Fig. 11-16) are designed to hold four cutting tools, which can be easily indexed for use as required. Several operations, such as turning, grooving, threading, and parting, may be performed on a workpiece by loosening the locking handle and rotating the holder until the desired toolbit is in the cutting position. This reduces the setup time for various toolbits, thereby increasing production.

QUICK-CHANGE TOOLPOST AND HOLDERS

Quick-change toolholders are made in different styles to accommodate different types of cutting tools. Each holder is dovetailed and fits on a dovetailed toolpost (Fig. 11-17A), which is mounted on the compound rest.

Courtesy Monarch Machine Tool Company

Fig. 11-16
A turret toolpost can hold up to four cutting tools and can be indexed quickly for different operations.

The tool is held in position by setscrews and is generally sharpened in the holder, after which it is preset to a gauge. After a tool becomes dull, the unit (the holder and tool) may be replaced with another preset unit. Each toolholder (Fig. 11-17B) fits onto the dovetail on the toolpost and is locked in position by means of a clamp. A knurled nut on each holder provides for vertical adjustment of the unit.

Courtesy DoAll Company

Fig. 11-17A
The dovetailed toolpost permits tools to be changed quickly and accurately.

Fig. 11-17B
Quick-change boring tools and holder.

WORK-HOLDING DEVICES

Work-holding devices are used to hold work in a lathe while cutting operations are performed. Lathe centres, chucks, face plates, mandrels, and steady rests are some of the common work-holding devices used for lathe work.

LATHE CENTRES

Most turning operations can be performed between the centres on a lathe. Centres are provided in a variety of types to suit the job required. Probably the most common centres used in school shops are the solid 60° centres with a Morse taper shank (Fig. 11-18A). These are usually made from a high-speed steel or from machine steel with carbide inserts or tips.

The *revolving* or *live tailstock centre* (Fig. 11-18B) is used to replace the standard solid dead centre for many applications. This type of centre contains precision anti-friction bearings to take both radial and axial thrusts. These centres are required when machining work at high speeds with carbide cutting tools, where the heat of friction causes the work to expand. The end or axial thrust created by the heated work is taken by the revolving centre and no adjustment is normally required. Since these centres require no centre lubricant, they are particularly useful when turning long shafts, where adjusting (and lubricating) the dead centre during the cutting operation would affect the finish of the workpiece.

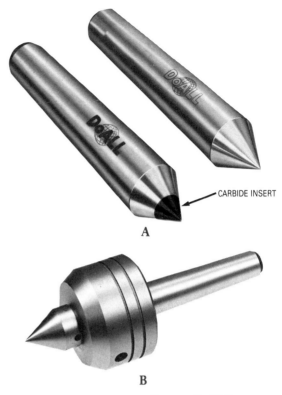

Courtesy DoAll Company

Fig. 11-18
Two styles of 60° centres with Morse taper shanks.

CHUCKS

The most commonly used chucks for lathe work are: 3-jaw universal, 4-jaw independent, combination, and collet.

The *3-jaw universal chuck* (Fig. 11-19) is used to hold round and hexagonal work. It grasps the work quickly to within a few hundredths of a millimetre or thousandths of an inch of accuracy, because the three jaws move simultaneously when adjusted by the chuck wrench. This simultaneous motion is accomplished by a *scroll plate* into which all three jaws fit. This type of chuck is made in various sizes from approximately 100 to 400 mm (4 to 16 in.) in diameter. It is usually provided with two sets of jaws, one for outside chucking and the other for inside chucking.

Fig. 11-19
The three-jaw universal chuck is used to hold round and hexagonal work.

The *4-jaw independent chuck* (Fig. 11-20) has four jaws, each of which can be adjusted independently by a square-end chuck wrench. It is used to hold round, square, hexagonal, and irregular-shaped workpieces. The jaws can be reversed to hold work by the inside diameter. Since each jaw can be adjusted independently, a workpiece can be adjusted to run absolutely true in a 4-jaw chuck.

A *combination chuck* has the mechanical features of both independent and universal chucks. The jaws can be operated individually by a screw, or universally by turning the adjusting socket which operates the bevel gear-driven scroll. This chuck is used for the same purposes as are the universal and independent chucks.

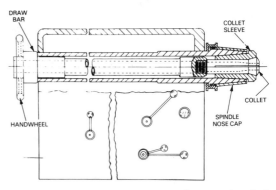

Courtesy Kostel Enterprises Ltd.

Fig. 11-21
A cross-sectional view of a headstock spindle, showing a draw-in collet assembly.

Courtesy Cushman Industries Inc.

Fig. 11-20
The four-jaw independent chuck is used to hold irregular-shaped workpieces.

The *collet chuck* (Fig. 11-21) is a draw-in chuck or collet that fits into the headstock spindle. The handwheel is used to tighten the collet on the work. Only true, round, and sized work can be held in a collet because of its spring limitations. There are collets for round, square, and hexagonal work.

The *Jacobs collet chuck* (Fig. 11-22) has a wider range than the draw-in collet chuck. Instead of a draw-in rod, an impact-tightening handwheel is used to close and release the collets on the workpiece. A set of 11 Rubber Flex collets, each having a range of almost 3 mm (1/8 in.), makes it possible to hold a wide range of work diameters.

Courtesy Jacobs Manufacturing Co.

Fig. 11-22
The Rubber-Flex Jacobs chuck has a wider range than most other collet chucks.

TYPES OF LATHE SPINDLES

There are two common types of lathe spindle noses upon which accessories such as drive plates and chucks are mounted. They are the tapered spindle nose and the cam-lock spindle nose. The threaded spindle nose may still be found on older lathes.

The *tapered spindle nose* (Fig. 11-23) had a 3-1/2 in. taper per foot. The chuck is fitted on the taper and positioned by the drive key. The threaded lock ring, fitted at the left end of the spindle nose, threads onto the chuck or drive plate and locks it in position. The *cam-lock spindle nose* (Fig. 11-24) has a very short taper (3 in. taper per foot). The chuck or drive plate is located by the taper on the spindle and held in position by three cam-lock devices.

Courtesy Kostel Enterprises Ltd.

Fig. 11-24
The type D-1 cam-lock spindle nose.

MANDRELS

A mandrel is used to hold an internally machined workpiece between centres so that further machining operations will be concentric with the bore. The mandrel (Fig. 11-25A), when pressed into a hole, allows the work to be mounted between centres and also provides a means of driving the workpiece (Fig. 11-25B).

Courtesy Kostel Enterprises Ltd.

Fig. 11-23
The type L spindle nose.

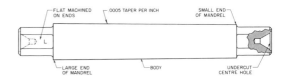

Fig. 11-25A
A plain mandrel.

WORK-HOLDING DEVICES 131

Courtesy South Bend Lathe, Inc.

Fig. 11-25B
A workpiece held on a mandrel for machining.

STEADY RESTS

A steady rest (Fig. 11-26) is used to support long work held in a chuck or between lathe centres. It is fastened to the lathe bed, and its three jaws are adjusted to lightly contact the work diameter and prevent the work from springing during a machining operation.

Courtesy Kostel Enterprises Ltd.

Fig. 11-26
A steady rest is used to support long, slender work for machining.

The follower rest (Fig. 11-27) is mounted on the saddle and supports the top and back of the work being turned. It prevents the work from springing up and away from the cutting tool when a cut is being taken on long work.

Courtesy South Bend Lathe, Inc.

Fig. 11-27
The follower rest travels along the work to prevent it from springing during machining.

ALIGNMENT OF LATHE CENTRES

To produce a parallel diameter on work machined between centres, it is important that the headstock and tailstock centres are in line. If the lathe centres are not aligned with each other, the diameter which is cut will be tapered (larger at one end than the other end). There are three common methods used to check the alignment of lathe centres:

1. By having the lines on the back of the tailstock in line. (This should always be done as a first step in aligning centres.)
2. By taking a light trial cut to the same depth setting near each end of the work and measuring the diameters with a micrometer.
3. By using a dial indicator and a parallel test bar. (This method achieves greatest accuracy.)

The lathe tailstock consists of two halves, the *baseplate* and the *tailstock body* (Fig. 11-28). The tailstock body can be adjusted either toward or away from the cutting tool by means of the two adjusting screws (G and F). This allows the tailstock centre to be aligned with the headstock centre.

Courtesy Kostel Enterprises Ltd.

Fig. 11-28
The tailstock may be adjusted for taper or parallel turning.

To Align Centres By the Tailstock Graduations

1. Loosen the tailstock clamp nut or lever.
2. Loosen one adjusting screw and tighten the other one depending upon the direction the tailstock must be moved.
3. Continue adjusting the screws until the line on the tailstock body matches the line on the baseplate (Fig. 11-28).
4. Tighten the loose adjusting screw to hold the tailstock body in position.
5. Recheck the tailstock graduations and then tighten the tailstock clamp nut or lever.
6. Mount the cutting tool on the left-hand side of the compound rest and on centre (Fig. 11-29).
7. Mount the workpiece between centres.
8. Set the lathe for the correct speed.
9. Take a light trial cut.

Courtesy Kostel Enterprises Ltd.

Fig. 11-29
The toolbit is set on centre and to the left of the compound rest.

10. Stop the machine, and check the diameters for taper.
11. If there is too much taper, follow steps 1 to 4 and readjust the tailstock in the proper direction.

To Align Centres With the Trial Cut Method

1. Take a trial cut from the work at the tailstock end (section A) deep enough to produce a true diameter (Fig. 11-30A). This cut should be about 6 mm (1/4 in.) long.

Courtesy Kostel Enterprises Ltd.

Fig. 11-30A
A light cleanup cut is taken for about 6 mm (1/4 in.) at end A.

ALIGNMENT OF LATHE CENTRES 133

Courtesy Kostel Enterprises Ltd.

Fig. 11-30B
A cut of the same depth is made at the other end of the work B.

2. Disengage the automatic feed, and note the reading on the graduated collar of the crossfeed screw.
3. Turn the crossfeed handle *counterclockwise* to bring the cutting tool away from the work.
4. Move the carriage until the cutting tool is about 25 mm (1 in.) from the lathe dog.
5. Start the lathe.
6. Turn the crossfeed handle *clockwise* until it is at the same graduated collar setting as it was at section A.
7. Cut section B about 12 mm (1/2 in.) long (Fig. 11-30B).

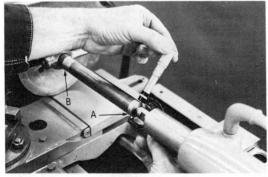

Courtesy Kostel Enterprises Ltd.

Fig. 11-31
Measuring the size after the trial cut.

8. Stop the lathe, and measure both diameters with a micrometer (Fig. 11-31).
9. If both diameters are the same, the lathe centres are in line. If the diameters are different, move the tailstock one-half the difference between the two diameters:
    (a) *Away from* the cutting tool if the diameter at the *tailstock end is smaller*.
    (b) *Towards* the cutting tool if the diameter at the *tailstock end is larger*.
10. If necessary, readjust the tailstock and take trial cuts until both diameters are the same.

Courtesy Kostel Enterprises Ltd.

Fig. 11-32
Mounting the dial indicator at centre height.

To Align Centres With a Test Bar and Dial Indicator

1. Clean the lathe centres and the centre holes in the test bar.
2. Mount a dial indicator, with the indicator plunger on centre and in a horizontal position, in the toolpost or on the lathe carriage (Fig. 11-32).
3. Adjust the test bar snugly between centres, and then tighten the tailstock spindle clamp.
4. Turn the crossfeed handle until the indicator registers approximately one-quarter of a revolution on section A,

Courtesy Kostel Enterprises Ltd.

Fig. 11-33
Setting the indicator bezel to "zero".

and set the indicator bezel to zero (Fig. 11-33).
5. Turn the carriage handwheel to bring the indicator into contact with section B (Fig. 11-34).
6. Compare the two indicator readings. If they are not the same, return the carriage until the indicator again registers on section A.
7. Loosen the tailstock clamp nut and, by means of the tailstock adjusting screws, move the tailstock in the proper direction the difference between the readings at sections A and B.

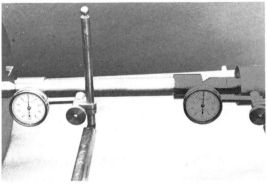

Courtesy Kostel Enterprises Ltd.

Fig. 11-34
If the readings at both ends are the same, the tailstock is properly aligned.

8. Tighten the tailstock clamp nut, and repeat steps 4 to 7 to recheck the centre alignment.

Mounting and Removing Lathe Centres

Lathe centres are removed to allow other operations to be done on a lathe, to clean the centres, to obtain centre trueness, and to replace worn centres.

Courtesy Kostel Enterprises Ltd.

Fig. 11-35
Removing the live centre using a knockout bar.

Figure 11-35 illustrates how a live centre is removed by using a *knockout bar* that is pushed through the headstock spindle. A slight tap is required to relieve the tapered centre. When removing the live centre with a knockout bar, place a cloth over the live centre and hold it with one hand to prevent an accident or damage to the centre. The dead centre can be removed by winding the tailstock spindle back into the tailstock. A screw knockout in the tailstock releases the dead centre.

Before replacing the centres and centre sleeve, clean them thoroughly, and make sure that all the cuttings, burrs, and oil are removed from the tapered portion of the headstock and tailstock spindles. Insert the centres in the headstock or tailstock with a sharp snap.

CENTRE LUBRICANTS

A centre lubricant such as special centre lube, white lead and oil, or red lead and oil must be used to lubricate the solid dead centre, to overcome the friction and heat caused by the revolving work.

MOUNTING WORK BETWEEN CENTRES

Work that is mounted between centres can be machined, removed, and set up for additional machining, and still maintain the same degree of accuracy. The cutting tool and the work must be properly set up, or damage to the machine, work, and lathe centres will result.

To Set Up the Cutting Tool

1. Move the toolpost to the left-hand side of the compound rest.
2. Mount a toolholder so that its setscrew is close to the toolpost.
   For the best rigidity, have about the width of a thumb between the toolholder screw and the toolpost (Fig. 11-36).
3. Set the toolholder so that it is at right angles to the work or pointing slightly towards the tailstock.
4. Insert the desired toolbit in the toolholder so that it extends no more than 12 mm (1/2 in.).
5. Use two-finger pressure on the wrench to tighten the toolholder setscrew.
   NOTE: *Tightening the setscrew too tightly may break the toolbit.*
6. Adjust the toolholder until the point of the cutting tool is even with the lathe centre point.
7. Tighten the *toolpost screw securely* to prevent the toolholder from moving under the pressure of the cut.

To Mount Work Between Centres Using a Solid Dead Centre

1. Check the live centre by holding a piece of chalk close to it while it is revolving.
   If the live centre is not running true, the chalk will mark only the high spot.
2. If the chalk marks the high spot, remove the live centre from the headstock and clean the tapers on the centre and the headstock spindle.
3. Replace the live centre and check again for trueness.

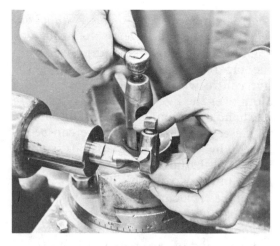

Courtesy Kostel Enterprises Ltd.

Fig. 11-36
Setting the cutting tool on centre and close to the toolpost.

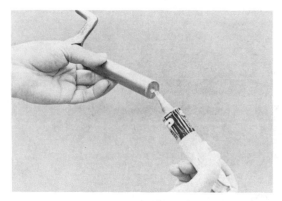

Courtesy Kostel Enterprises Ltd.

Fig. 11-37
The dog is placed loosely on one end, and centre lubricant is applied to the other.

Courtesy Kostel Enterprises Ltd.

Fig. 11-38
Securing the dog on the workpiece.

4. Adjust the tailstock spindle until it extends about 65 to 75 mm (2-1/2 to 3 in.) beyond the tailstock.
5. Loosen the tailstock clamp nut or lever.
6. Place the lathe dog on the end of the work with the bent tail pointing to the left (Fig. 11-37).
7. Lubricate the dead centre with a suitable centre lubricant.
8. Place the end of the work with the lathe dog on the live centre.
9. Slide the tailstock toward the headstock until the dead centre supports the other end of the work.
10. Tighten the tailstock clamp nut or lever.

Courtesy Kostel Enterprises Ltd.

Fig. 11-39
Adjusting the centre tension.

11. Adjust the tail of the dog in the slot of the drive plate and tighten the lathe dog screw (Fig. 11-38).
    NOTE: *Be sure that the tail of the dog does not bind in the drive plate slot.*
12. Turn the drive plate by hand until the slot and lathe dog are parallel with the bed of the lathe.
13. Hold the tail of the dog up in the slot and tighten the tailstock handwheel only enough to hold the lathe dog in the up position (Fig. 11-39).
14. Turn the tailstock handwheel backwards only until the lathe dog drops in the drive plate slot.
15. Hold the tailstock handwheel in this position and tighten the tailstock spindle clamp with the other hand.
16. Check the centre adjustment by attempting to move the work endwise. *There should be no end play.*

Courtesy Kostel Enterprises Ltd.

Fig. 11-40
Checking the centre tension.

17. Check the centre tension by raising the tail of the dog to the top of the faceplate slot (Fig. 11-40), and remove your hand. The tail of the dog should drop of its own weight.
18. Move the carriage to the furthest position (left-hand end) of cut, and revolve the lathe spindle by hand to see that the dog does not hit the compound rest.

To Mount Work Between Centres Using a Revolving Tailstock Centre

The use of a revolving or live dead centre eliminates the need for centre lubrication when turning a long workpiece. It is also useful for heavy duty and high-speed turning. When using this type of centre proceed as follows.

Follow steps 1 to 9, omitting step 7, in the previous section, "To Mount Work Between Centres Using a Solid Dead Centre".

10. Tighten the tailstock handwheel until the workpiece is held snugly between centres.
11. Tighten the tailstock spindle clamp.
12. Re-check the set-up to be sure that the dog does not bind in the slot of the drive plate.

NOTE: When end-facing work using this type of centre, always recess the outer end of the centre hole to prevent damaging the revolving centre with the cutting tool.

MACHINING SEQUENCE

Proper procedure should be followed when machining any part so that it can be machined accurately and as quickly as possible. It would be impossible to list the exact sequence of operations that should be followed for every type of workpiece machined on a lathe. The following guidelines, with minor exceptions, would apply to work being machined between centres or in a chuck.

General Rule for Machining Round Work

1. Rough-turn all diameters and lengths to within 0.80 mm (1/32 in.) of the size required.
2. Machine the largest diameter first and progress to the smallest.
3. Special operations, such as knurling or grooving, should be done next.
4. Cool the workpiece before starting any finishing operations.
5. Finish turn all diameters and lengths. Always start with the largest diameter and work to the smallest.

MACHINING BETWEEN CENTRES

There are times when it is necessary to machine the entire length of a round workpiece. The workpiece shown in Fig. 11-41 is a typical part which can be machined between the centres on a lathe.

Machining Sequence

1. Cut off a piece of steel 1/8 in. (3.18 mm) longer and 1/8 in. (3.18 mm) larger in diameter than required.
   In this case the diameter of the steel cut off would be 1-5/8 in. (41.3 mm) and its length would be 16-1/8 in. (409.6 mm).
2. Hold the work in a three-jaw chuck, face one end square, and then drill the centre hole.
3. Face the other end to length, and then drill the centre hole.
4. Mount the workpiece between the centres on a lathe.

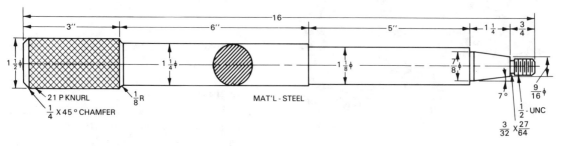

Fig. 11-41
The print of a round workpiece to be machined.

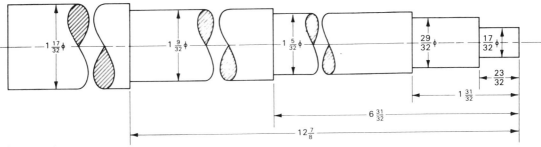

Fig. 11-42
The rough-turned diameters and the lengths on the work.

5. Rough-turn the largest diameter to within 1/32 in. (0.80 mm) of finish size or 1-17/32 in. (38.9 mm).

   The purpose of the rough cut is to remove excess metal as quickly as possible.

6. Finish-turn the diameter to be knurled.

   The purpose of the finish cut is to cut work to the required size and produce a good surface finish.

7. Knurl the 1-1/2 in. (38.1 mm) diameter.
8. Machine the 45° chamfer on the end.
9. Reverse the work in the lathe, being sure to protect the knurl from the lathe dog with a piece of soft metal.
10. Rough turn the 1-1/4 in. (31.8 mm) diameter to 1-9/32 in. (32.5 mm).

    Be sure to leave the length of this section 1/8 in. (3.18 mm) short [12-7/8 in. (327 mm) from the end] to allow for finishing the 1/8 in. (3.18 mm) radius.

11. Rough-turn the 1-1/8 in. (28.6 mm) diameter to 1-5/32 in. (29.4 mm) (Fig. 11-42).

    Leave the length of this section 1/32 in. (0.80 mm) short [6-31/32 in. (177 mm) from the end] to allow for finishing the shoulder.

12. Rough-turn the 7/8 in. (22.2 mm) diameter to 29/32 in. (23.0 mm) for a distance of 1-31/32 in. (50.0 mm) from the end.
13. Rough-turn the 1/2 in. (12.7 mm) diameter to 17/32 in. (13.5 mm) for a distance of 23/32 in. (18.2 mm).
14. Cool the work to room temperature before starting the finishing operations.
15. Finish-turn the 1-1/4 in. (31.8 mm) diameter to 12-7/8 in. (327 mm) from the end.
16. Mount a 1/8 in. (3.18 mm) radius tool, and finish the corner to the correct length (Fig. 11-43).

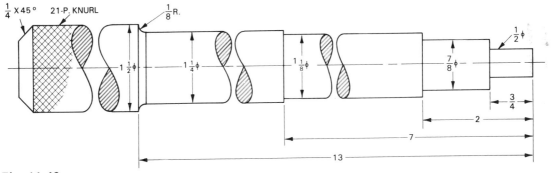

Fig. 11-43
The workpiece turned to finished diameters and lengths.

MACHINING SEQUENCE

17. Finish-turn the 1-1/8 in. (28.6 mm) diameter to 7 in. (178 mm) from the end.
18. Finish-turn the 7/8 in. (22.2 mm) diameter to 2 in. (50.8 mm) from the end.
19. Set the compound rest to 7°, and machine the taper to size.
20. Finish-turn the 1/2 in. (12.7 mm) diameter to 3/4 in. (19.0 mm) from the end.
21. With a cut-off tool cut the groove at the end of the 1/2 in. (12.7 mm) diameter (Fig. 11-44).
22. Chamfer the end of the section to be threaded.
23. Set the lathe for threading, and cut the thread to size.

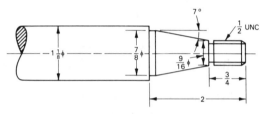

Fig. 11-44
Special operations completed on the part.

FACING WORK BETWEEN CENTRES

Facing is a squaring operation performed on the ends of work after it has been cut off by a saw. In order to produce a flat surface when facing between centres, the lathe centres must be in line. The purposes of facing are:
(a) to provide a true flat surface, square with the axis of the work;
(b) to make a smooth surface from which to take measurements;
(c) to cut work to a required length.

To Face Work Between Centres
1. Set the toolholder to the left-hand side of the compound rest.
2. Mount a facing tool in the toolholder, having it extend only about 12 mm (1/2 in.), and tighten the toolholder setscrew using only two-finger pull.
3. Adjust the toolholder until the point of the facing tool is at the same height as the lathe centre point, then tighten the toolpost lightly (Fig. 11-45).

Courtesy Kostel Enterprises Ltd.

Fig. 11-45
The toolholder is gripped short, and the toolbit is set on centre.

4. Tap the toolholder lightly until the point of the cutting tool is closest to the work and there is a space along the side (Fig. 11-46).
5. Tighten the toolpost securely to keep the toolholder in position.
6. If the entire end must be faced, insert a half centre in the tailstock spindle (Fig. 11-46).
7. Bring the toolbit to the end of the work as in Fig. 11-46 by moving the carriage with the apron handwheel.
8. Set the lathe to the correct speed for the material being cut.
9. Move the toolbit in as close to the centre of the work as possible, and set the depth of cut by using the apron handwheel.
10. Tighten the carriage lock, using only a two-finger pull on the wrench (Fig. 11-47).

Courtesy Kostel Enterprises Ltd.

Fig. 11-46
A half-centre allows the entire end to be faced.

Courtesy Kostel Enterprises Ltd.

Fig. 11-47
When locking the carriage, use only a two-finger pull.

11. Feed the tool out by turning the cross-feed screw handle slowly and steadily with two hands, cutting from the centre outwards.

12. Repeat operations 9 to 11 until the work is the correct length. (Before facing, mark the length with a centre punch mark and then face until the punch mark is cut in half.)

   When facing, all cuts must begin at the centre and feed toward the outside.

13. Check the surface for flatness using the edge of a rule (Fig. 11-48).

   If the centre is high or low, recheck the alignment of the lathe centres and reface the work.

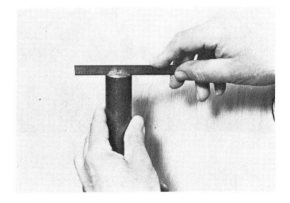

Courtesy Kostel Enterprises Ltd.

Fig. 11-48
The surface is flat if no light shows between the work and the rule.

PARALLEL TURNING

Work is generally machined on a lathe for two reasons: to cut it to size and to produce a true diameter. Work that must be cut to size and also be the same diameter along the entire length of the workpiece involves the operation of parallel turning. In order to produce a parallel diameter, the headstock and tailstock centres must be in line. Many factors determine the amount of material which can be removed on a lathe at one time. However, whenever possible, work should be cut to size in two cuts: a roughing cut and a finishing cut.

Courtesy South Bend Lathe, Inc.

Fig. 11-49
Micrometer collars assist the operator in setting the correct depth of cut.

GRADUATED MICROMETER COLLARS

Graduated micrometer collars are sleeves or bushings that are mounted on the compound rest and crossfeed screws (Fig. 11-49). They assist the lathe operator to set the cutting tool accurately to remove the required amount of material from the workpiece. The micrometer collars on lathes using the inch system of measurement are usually graduated in thousandths of an inch (0.001 in.). The collars on lathes using the metric system of measurement are usually graduated in fiftieths of a millimetre (0.02 mm).

The graduated collar indicates only the distance that the cutting tool has been moved toward the work. Therefore on any machines where the work revolves (lathes, cylindrical grinders, boring mills, etc.), the cutting tool should be set in *only half the amount of metal to be removed* because material is removed from the circumference of the work piece.

INCH LATHES

The circumference of the crossfeed and compound rest screw collars on lathes using the inch system of measurement is usually divided into 100 or 125 equal divisions, each having a value of 0.001 in. Therefore, if the crossfeed screw is turned *clockwise* 20 graduations, the cutting tool will be moved 0.020 in. towards the work. Because the work in a lathe revolves, a 0.020 in. depth of cut will be taken from the entire work circumference, thereby reducing the diameter 0.040 in. (2 × 0.020 in.) (Fig. 11-50).

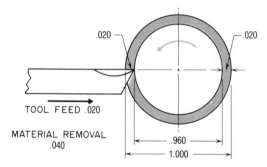

Fig. 11-50
Since the work in a lathe revolves, a 0.020 in. depth of cut will reduce the diameter by 0.040 in.

METRIC LATHES

The circumference of the crossfeed and compound rest screw collars on lathes using the metric system of measurement is usually divided into 200 or 250 equal divisions, each having a value of 0.02 mm. Therefore if the crossfeed screw is turned *clockwise* 30 graduations, the cutting tool will be moved 30 × 0.02 mm or 0.6 mm toward the work. Because the work in a lathe revolves, a 0.6 mm depth of cut will remove 1.2 mm from the diameter of a workpiece.

To Set an Accurate Depth of Cut

1. Move the toolpost to the left-hand side of the compound rest.
2. Set the cutting tool to centre, and tighten the toolpost screw *securely*.
3. Start the lathe and move the carriage until the toolbit overlaps the right-hand end of the workpiece by approximately 1.5 mm (1/16 in.).
4. Feed the toolbit in with the crossfeed handle until a light cut is made around the entire circumference of the work.
5. Turn the carriage handwheel until the toolbit just clears the right-hand end of the work.
6. Turn the crossfeed handle clockwise 0.12 mm (or 0.005 in. for inch lathes) and set the graduated collar to zero without moving the crossfeed handle.
7. Take a trial cut about 6 mm (1/4 in.) long (Fig. 11-51).
   The purpose of this trial cut is to:
   (a) produce a true diameter on the work;
   (b) set the cutting tool to the diameter;
   (c) set the crossfeed graduated collar to the diameter.
8. Stop the lathe and *be sure that the crossfeed handle setting is not moved*.
9. Turn the carriage handwheel until the toolbit clears the right-hand end of the work.
10. Measure the diameter of the trial cut with a micrometer (Fig. 11-52), and then calculate the amount of metal yet to be removed.

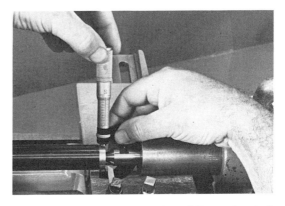

Courtesy Kostel Enterprises Ltd.

Fig. 11-52
Measuring the diameter of a trial cut with a micrometer.

11. Turn the crossfeed handle clockwise until the graduated collar moves *one-half the amount of material to be removed*.
    For example, if 0.50 mm (0.020 in.) must be removed, the crossfeed handle should be set in 0.25 mm (0.010 in.).
12. Take another trial cut 6 mm (1/4 in.) long and measure the diameter with a micrometer.
13. Adjust the crossfeed handle setting if necessary.

ROUGH TURNING

Rough turning is used to remove most of the excess material as quickly as possible and to true the work diameter. The roughing cut should be taken to within 0.80 mm (1/32 in.) of the finished size of the workpiece. Generally one roughing cut should be taken if up to 12 mm (1/2 in.) is to be removed from the diameter.

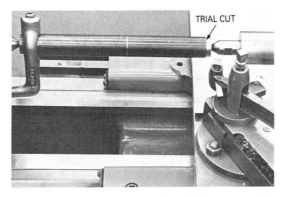

Courtesy Kostel Enterprises Ltd.

Fig. 11-51
A light trial cut should be made, to establish a size, before turning a diameter.

To Rough Turn a Diameter

1. Mount a rough turning or general purpose toolholder in the toolpost.
2. Set the toolpost on the left side of the compound rest.
3. Have the toolholder extend as little as possible beyond the toolpost, and set the point of the cutting tool even with the lathe centre point (Fig. 11-53).

Courtesy Kostel Enterprises Ltd.

Fig. 11-53
The toolholder should be set as close as possible (about the width of a thumb) to the toolpost and the point of the toolbit on centre.

4. Adjust the toolholder so that it is pointing slightly toward the tailstock (Fig. 11-54A). An incorrectly set toolholder is shown in Fig. 11-54B.
5. Tighten the toolpost screw *securely*.
6. Set the lathe speed for the material being cut. (See Table 11-1.)
7. Set the quick-change gearbox (Fig. 11-55) for the rough cut feed (generally about 0.25 to 0.50 mm (or 0.010 to 0.020 in. for inch lathes).
8. Take a light trial cut about 6 mm (1/4 in.) long at the right-hand end of the work.
9. Stop the lathe, but *do not move the crossfeed handle setting or the graduated collar.*
10. Turn the carriage handwheel until the cutting tool clears the right-hand end of the work.
11. Measure the diameter of the trial cut, and calculate how much material must be removed.

Always leave the rough cut diameter from 0.80 to 1.30 mm (or 0.030 to 0.050 in.) over the finished size required. This will leave enough material for the finish cut (Fig. 11-56).

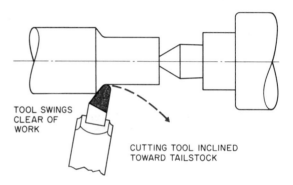

Fig. 11-54A
If a correctly set toolholder moves under the pressure of a cut, the cutting tool will swing away from the work.

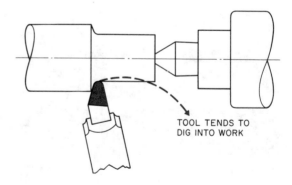

Fig. 11-54B
If an incorrectly set toolbar moves under pressure of a cut, the work will be machined undersize and probably scrapped.

Courtesy Kostel Enterprises Ltd.

Fig. 11-55
Set the levers to the positions indicated on the quick-change gearbox chart for rough and finish cuts.

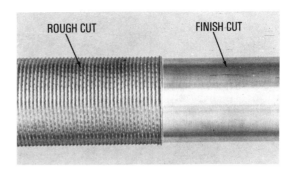

Courtesy Kostel Enterprises Ltd.

Fig. 11-56
The surface finishes produced by rough and finish cuts.

12. Turn the crossfeed handle *clockwise* one-half the calculated amount and take a trial cut 6 mm (1/4 in.) long.
13. Measure the diameter with a micrometer, and reset the depth of cut if necessary.
14. Turn the rough cut to the required length.

FINISH TURNING

The purpose of finish turning is to bring the workpiece to the required size and to produce a good surface finish. Generally only one finish cut is required since no more than 0.80 to 1.30 mm (or 0.030 to 0.050 in.) should be left on the diameter for the finish cut. The toolbit should have a slight radius on the point, and the lathe should be set for a 0.07 to 0.12 mm (or 0.003 to 0.005 in.) feed. Be sure that the lathe centres are aligned exactly; otherwise a taper will be cut on the workpiece.

To Finish Turn a Diameter

1. Set the lathe speed for finish turning (Table 11-1).
2. Set the quick-change gearbox for the finish feed at approximately 0.07 to 0.12 mm or 0.003 to 0.005 in. for inch lathes.
3. Take a *light trial cut* 6 mm (1/4 in.) long from the diameter at the tailstock end.
4. Disengage the feed and stop the lathe, but *do not move the crossfeed handle setting*.
5. Turn the carriage handwheel until the cutting tool clears the right-hand end of the work.
6. Measure the diameter with a micrometer, and calculate the amount of material that must still be removed.
7. Turn the crossfeed handle *clockwise* one-half the calculated amount (the difference between the trial cut size and the finished diameter), and take a trial cut 6 mm (1/4 in.) long.
8. Stop the lathe and measure the diameter.
9. Reset the crossfeed handle setting if necessary and turn the finish cut to the required length.

To Check a Diameter With an Outside Caliper

Whenever possible, a diameter should be measured with a micrometer; however, there may be times when a micrometer is not available or it is necessary to measure narrow grooves. Diameters may be measured within a reasonable degree of accuracy with outside calipers if they are properly set and used.

PARALLEL TURNING 145

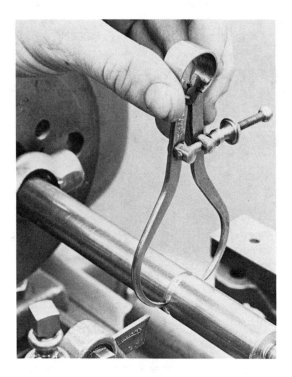

Courtesy Kostel Enterprises Ltd.

Fig. 11-57
Check the diameter of the groove with an outside caliper only when the machine has stopped.

1. Stop the machine.

   NOTE: *Never measure work that is revolving — it is dangerous and the measurement taken will be inaccurate.*

2. Turn the adjusting screw until the legs of the caliper lightly contact the centre of the diameter.
3. Check the setting by holding the spring of the caliper between the thumb and index finger and allowing the legs to drop over the diameter (Fig. 11-57).
   If the setting is correct, the caliper should just drop over the diameter by its own weight. *Do not force a caliper over a diameter.*
4. Adjust the caliper until just a *slight drag* is felt as the caliper legs pass over the diameter.
5. Use a rule to measure the distance between the two caliper legs.
   This method of measurement is only as accurate as a rule can be read. If greater accuracy is required when measuring a groove, a knife-edge vernier caliper should be used.

SHOULDER TURNING

Whenever more than one diameter is machined on a shaft, the section joining each diameter is called a shoulder or step. The square, filleted, and chamfered shoulders are most commonly used in machine shop work (Fig. 11-58).

To Machine a Square Shoulder

1. Lay out the length of the shoulder with a centre punch mark or cut a light

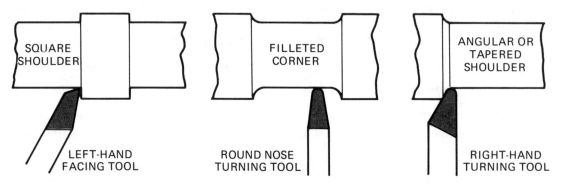

Fig. 11-58
Three common types of shoulders and the turning tools used for each.

groove at this point with a sharp toolbit (Fig. 11-59 A, B).
2. Rough and finish turn the small diameter to within 0.80 mm (1/32 in.) of the required length.
3. Mount a facing tool and set it for the facing operation.
4. Start the lathe and feed the cutting tool in until it lightly marks the small diameter near the shoulder.

A

B

Courtesy Kostel Enterprises Ltd.

Fig. 11-59
Marking the length of a shoulder: (A) with a centre punch mark; (B) by cutting a light groove around the workpiece.

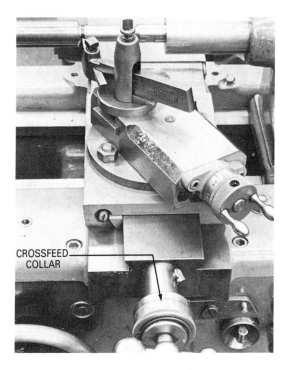

Courtesy Kostel Enterprises Ltd.

Fig. 11-60
The cutting tool can be returned to the same position each time if the graduated collar setting is used.

5. Note the reading on the crossfeed graduated collar (Fig. 11-60).
6. Turn the carriage handwheel to start a light cut.
7. Face the shoulder by turning the crossfeed handle *counterclockwise*.
8. Return the crossfeed handle to the original graduated collar setting.
9. Repeat steps 6 and 7 until the shoulder is to the correct length.

To Machine a Filleted Shoulder
1. Lay out the length of the shoulder with a centre punch mark or by cutting a light groove at this point. (Fig. 11-59A, B).
2. Rough and finish turn the small diameter to the correct length *minus the radius to be cut.*

SHOULDER TURNING 147

Courtesy Kostel Enterprises Ltd.

Fig. 11-61
A radius tool is used to produce a filleted shoulder.

For example, a 76 mm (3 in.) length with a 3 mm (1/8 in.) radius should be turned 72 mm (2-7/8 in.) long.

3. Mount the correct radius toolbit and set it to centre (Fig. 11-61).
4. Set the lathe for one-half the turning speed.
5. Start the lathe and feed the cutting tool in until it *lightly marks* the small diameter near the shoulder.
6. Slowly feed the cutting tool sideways with the carriage handwheel until the shoulder is cut to the correct length.

Courtesy Kostel Enterprises Ltd.

Fig. 11-62
Using a protractor to set the cutting edge of the toolbit to 45°.

To Machine Angular Shoulders

Long angular shoulders are generally produced by swivelling the compound rest to the required angle and then feeding the compound rest screw. Short angular shoulders can be cut by setting the side of the toolbit to the required angle (Fig. 11-62) and feeding it against the workpiece (Fig. 11-63).

Courtesy Kostel Enterprises Ltd.

Fig. 11-63
Machining a bevelled shoulder with the side of the toolbit.

FILING IN A LATHE

Filing in a lathe is used to remove burrs, tool marks, and sharp corners. It is not considered good practice, however, to file a diameter to size, for too much filing will tend to produce a diameter which is out of round. The National Safety Council recommends grasping the file handle with the left hand, so that arms and hands can be kept clear of the headstock.

When filing or polishing in a lathe, it is good practice to cover the lathe bed with a piece of paper to prevent filings from getting into the slides and causing excessive wear and damage to the lathe. A cloth is not advisable for this purpose because of the danger of having it caught in the revolving work or the lathe.

NOTE: Before starting to file, it is wise to observe the following:

(a) Always remove watches and rings.

(b) Roll coat and shirt sleeves up above the elbows.

(c) Never use a file without a properly fitted handle.

To File Work in a Lathe

1. Cover the lathe bed with paper (Fig. 11-64).
2. Set the lathe at twice the speed used for turning.
3. Adjust the work freely between centres. (Use a rotating dead centre if it is available.)
4. Disengage the lead screw by placing the reverse lever in a neutral position.
5. Select a suitable long angle lathe or mill file.
6. Grip the lathe file handle in the left hand, using the fingers of the right hand to balance and guide the file at the point (Fig. 11-65).
7. Move the file along the work after each stroke, so that each cut overlaps approximately one-half of the width of the file.

Courtesy Kostel Enterprises Ltd.

Fig. 11-64
Covering the lathe bed with paper will prevent damage to the bed by small metal chips produced by filing.

Courtesy Kostel Enterprises Ltd.

Fig. 11-65
The left-hand method of filing is safer.

8. Use long strokes, and apply pressure only on the forward stroke.
9. Use approximately 35 to 40 strokes/min.
10. If the file loads up with cuttings, clean it with a file brush and rub a little chalk on the file teeth to reduce pinning.

POLISHING

Polishing is a finishing operation which generally follows filing to improve the surface finish on the work. The finish obtained on the diameter is directly related to the coarseness of the abrasive cloth used. A fine grit abrasive cloth produces the best surface finish. Aluminum oxide abrasive cloth should be used for polishing most ferrous metals, while silicon carbide abrasive cloth is used on nonferrous metals.

To Polish Work in a Lathe

1. Be sure that all loose clothing is tucked in to prevent it from becoming caught by the revolving work.
2. Cover the lathe bed with paper (Fig. 11-64).
3. Set the lathe at a high speed, and disengage the lead screw and feed rod.

Fig. 11-66
A high surface finish can be produced with abrasive cloth.

Courtesy J.H. Williams & Co.

Fig. 11-67
Coarse, medium, and fine diamond and straight pattern knurling tools.

4. Mount work between centres freely with very little end play, or use a rotating dead centre.
5. Use a piece of 80 to 100 grit abrasive cloth about 25 mm (1 in.) wide for rough polishing.
6. Hold the abrasive cloth as shown in Fig. 11-66 to prevent the top end of the abrasive cloth from wrapping around the work and injuring the fingers.
7. Hold the long end of the abrasive cloth *securely* with one hand while the fingers of the other hand press the cloth against the diameter (Fig. 11-66).
8. Slowly move the abrasive cloth back and forth along the diameter to be polished.
9. Use a piece of 120 to 180 grit abrasive cloth for finish polishing.
10. Apply a few drops of oil to the abrasive cloth for the final passes along the diameter.

KNURLING

Knurling is a process of impressing diamond-shaped or straight indentations on the surface of work. The purposes of knurling are to improve the appearance of the work and to provide a better grip. It is done by forcing a knurling tool containing a set of hardened cylindrical patterned rolls against the surface of revolving work. Diamond and straight pattern rolls in three styles (fine, medium, and coarse) are illustrated in Fig. 11-67.

Knurling tools have a heat-treated body which is held in the toolpost and a set of hardened rolls mounted in a movable head. The knurling tool shown in Fig. 11-68A contains one set of rolls mounted in a self-centring head. The knurling tool in Fig. 11-68B contains three sets of rolls (fine, medium, and coarse) mounted in a revolving head which pivots on a hardened steel pin.

Fig. 11-68A
A knurling tool with one set of rolls in a self-centring head.

Fig. 11-68B
A knurling tool with three sets of rolls (coarse, medium, and fine) in a revolving head.

To Knurl a Diamond Pattern

1. Mount the work between centres with the required length of knurled section marked on the work.
   Use a revolving centre in the tailstock if it is available.
2. Set the lathe at 1/4 the speed used for turning.
3. Set the quick-change gearbox for a feed of 0.25 to 0.50 mm (0.010 to 0.020 in.).
4. Set the centre of the floating head of the knurling tool even with the dead centre of the lathe (Fig. 11-69).
5. Adjust the knurling tool so that it is at right angles to the work (Fig. 11-70).
6. Tighten the toolpost screw *securely* so that the knurling tool will not move during the knurling operation.

Courtesy Kostel Enterprises Ltd.

Fig. 11-70
The knurling tool should be set at 90° to the work.

7. Set the knurling tool near the end of the work so that only 1/2 to 3/4 of the width of the knurling roll is on the work (Fig. 11-71).
   This generally results in easier starting and a better knurling pattern.
8. Force the knurling tool into the work approximately 0.65 mm (0.025 in.) and start the lathe.

Courtesy Kostel Enterprises Ltd.

Fig. 11-69
Setting a knurling tool to centre.

Courtesy Kostel Enterprises Ltd.

Fig. 11-71
The knurling tool is set so that about 1/2 to 3/4 of the roll width is on the work.

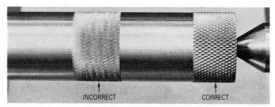

Courtesy Kostel Enterprises Ltd.

Fig. 11-72
Correct and incorrect knurling patterns.

OR

Start the lathe and then force the knurling tool into the work until the diamond pattern comes to a point.

9. Stop the lathe and examine the pattern. If necessary, reset the knurling tool.
   (a) If the pattern is incorrect (Fig. 11-72), it is usually because the knurling tool is not set on centre.
   (b) If the knurling tool is on centre and the pattern is not correct, it is generally due to worn knurling rolls. In this case, it will be necessary to set the knurling tool off square slightly so that the corner of the knurling rolls can start the pattern.
10. Once the pattern is correct, engage the automatic carriage feed and apply cutting fluid to the knurling rolls.

Courtesy Kostel Enterprises Ltd.

Fig. 11-73
Disengaging the automatic feed, during the knurling operation, will damage the knurling pattern.

11. Knurl to the proper length.
    Do not disengage the feed until the full length has been knurled; otherwise rings will form on the knurled pattern (Fig. 11-73).
12. If the knurling pattern is not to a point, reverse the lathe feed and take another pass across the work.

GROOVING

Grooving is an operation often referred to as recessing, undercutting, or necking. It is often done at the end of a thread, at the side of a shoulder, or for appearance. Grooves may be any desired shape, but they are generally square, round, or V-shaped (Fig. 11-74).

To Cut a Groove
1. Lay out the location of the groove, using a centre punch and layout tools.
2. Set the lathe at 1/2 the turning speed.
3. Mount the proper shaped toolbit in the toolholder and set the cutting tool to centre.
4. Locate the toolbit on the work at the position where the groove is to be cut.
5. Start the lathe and feed the cutting tool toward the work, using the crossfeed handle, until the toolbit lightly marks the work.
6. Hold the crossfeed handle in position and then set the graduated collar to zero.
7. Calculate how far the crossfeed screw must be turned to cut the groove to the proper depth.
8. Apply cutting fluid frequently, and groove the work to the proper depth.
9. Stop the lathe and check the depth of the groove with outside calipers.
10. It is desirable to move the carriage by hand a little to the right and left while grooving to overcome chatter.

NOTE: Wear safety goggles when grooving work in a lathe.

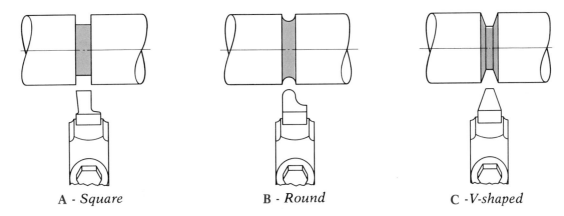

A - Square   B - Round   C - V-shaped

Fig. 11-74
Three types of common grooves.

TAPERS

A taper may be defined as a uniform increase or decrease in the diameter of a piece of work measured along its length. Metric tapers are expressed as a ratio of 1 mm per unit of length; for example 1:20 taper would have a 1 mm change in diameter in 20 mm of length. Tapers in the inch system are expressed in taper per foot or taper per inch. A taper provides a rapid and accurate method of aligning machine parts and an easy method of holding tools such as twist drills, lathe centres, and reamers. The American Standards Association classifies tapers used on machines and tools as *self-holding tapers* and *self-releasing* or steep tapers.

Self-holding tapers are those which remain in position due to the wedging action of the taper. The inch tapers of this series are composed of the Morse, Brown and Sharpe, and 3/4 in./ft. Machine tapers (Table 11-3). *Steep tapers*, such as those used on milling machine arbors and accessories, are held in the machine by a draw-bolt and are driven by lugs or keys.

Inch Tapers

Some of the tapers included in Table 11-3 are taken from the Morse, and Brown and Sharpe series. The following describes these and others which are sometimes used in machine shop work.

1. *Morse taper*, approximately 5/8 in. taper/ft., is a standard taper used for twist drills, reamers, end mills, and lathe centre shanks. The Morse taper has eight standard sizes from 0 to 7.
2. *Brown and Sharpe taper*, 1/2 in. taper/ft., is a standard taper used in all Brown and Sharpe machines, cutters, and drive shanks.
3. *Jarno taper*, 0.600 in. taper/ft., is used for some machine spindles.
4. *Standard taper pin*, 1/4 in. taper/ft., is a standard for all tapered pins used in the fabrication of machinery. They are listed by numbers from 0 to 10.
5. *Standard Milling Machine taper*, 3-1/2 in. taper/ft., is a self-releasing taper used exclusively on milling machine spindles and equipment.

Metric Tapers

Metric tapers are expressed as a ratio of one millimetre per unit of length. In Fig. 11-75 the work would taper one millimetre in a distance of twenty millimetres. This taper would then be expressed as a ratio of 1:20 and would be indicated on a drawing as Taper = 1:20.

Since the work tapers 1 mm in 20 mm of length, the diameter at a point 20 millimetres from the small diameter (d) will be 1 mm larger (d + 1).

Some common metric tapers are:
Milling machine spindle — 1:3.43
Morse taper shank — approximately 1:20
Tapered pins and pipe threads — 1:50

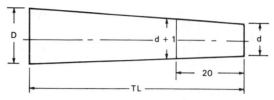

Fig. 11-75
The characteristics of a metric taper.

TABLE 11-3 BASIC DIMENSIONS OF SELF-HOLDING TAPERS

Number of Taper	Taper per Foot	Diameter at Gauge Line (A)	Diameter at Small End (D)	Length (P)	Series Origin
1	0.502	0.2392	0.200	15/16	
2	0.502	0.2997	0.250	1-3/16	Brown and Sharpe
3	0.502	0.3752	0.3125	1-1/2	Taper Series
* 0	0.624	0.3561	0.252	2	
1	0.5986	0.475	0.369	2-1/8	
2	0.5994	0.700	0.572	2-9/16	
3	0.6023	0.938	0.778	3-3/16	
4	0.6233	1.231	1.020	4-1/16	Morse Taper
4-1/2	0.624	1.500	1.266	4-1/2	Series
5	0.6315	1.748	1.475	5-3/16	
6	0.6256	2.494	2.116	7-1/4	
7	0.624	3.270	2.750	10	
200	0.750	2.000	1.703	4-3/4	
250	0.750	2.500	2.156	5-1/2	
300	0.750	3.000	2.609	6-1/4	
350	0.750	3.500	3.063	7	
400	0.750	4.000	3.516	7-3/4	
450	0.750	4.500	3.969	8-1/2	3/4 in. Taper
500	0.750	5.000	4.422	9-1/4	per Foot Series
600	0.750	6.000	5.328	10-3/4	
800	0.750	8.000	7.141	13-3/4	
1000	0.750	10.000	8.953	16-3/4	
1200	0.750	12.000	10.766	19-3/4	

* Taper #0 is not a part of the self-holding taper series. It has been added to complete the Morse taper series.

INCH TAPER CALCULATIONS

Most inch tapers cut on workpieces are expressed in taper per foot or degrees. If this information is not supplied, it is generally necessary to calculate the taper per foot of the workpiece. Taper per foot is the amount of difference between the large diameter and the small diameter of the taper in 12 in. of length. For example, if the tapered section on a piece of work is 12 in. long and the large diameter is 1 in. and the small diameter is 1/2 in., the taper per foot would be the difference between the large and small diameters, or 1/2 in. The main parts of an inch taper are: the amount of taper, the length of the tapered part, the large diameter, and the small diameter (Fig. 11-76).

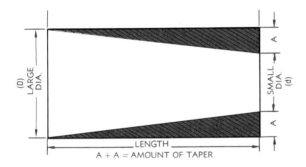

Fig. 11-76
The parts of an inch taper.

Since not all tapers are 12 in. long, if the small diameter, large diameter, and length of the tapered section are known, then the taper per foot can be calculated by applying the following formula:

$$\text{taper/ft.} = \frac{(D - d) \times 12}{T.L.}$$

where:
D = diameter at the large end of the taper
d = diameter at the small end of the taper
$T.L.$ = total length of the tapered section

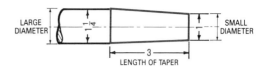

Fig. 11-77
The dimensions of an inch tapered workpiece.

EXAMPLE: To calculate the taper per foot of the workpiece shown in Fig. 11-77:

$$\text{taper/ft.} = \frac{(1\text{-}1/4 - 1) \times 12}{3}$$

$$= \frac{1/4 \times 12}{3}$$

$$= 1 \text{ in.}$$

If taper per inch is required, divide taper per foot by 12. For example, the 1 in. taper/ft. of the previous problem would have 0.083 taper/in. (1 in. ÷ 12 in.).

After the taper per foot has been calculated, no further calculations are necessary if the taper is to be cut with a taper attachment. If the taper is to be cut by the tailstock offset method, the amount to offset the tailstock must be calculated.

METRIC TAPER CALCULATIONS

If the small diameter (d), the unit length of taper (k), and the total length of taper (ℓ) are known, the large diameter (D) may be calculated.

In Fig. 11-78A, the large diameter (D) will be equal to the small diameter plus the amount of taper. The amount of taper for the unit length (k) is $(d + 1) - (d)$, or 1 mm.

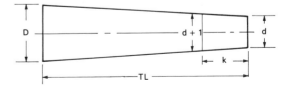

Fig. 11-78A
The dimensions of a metric taper.

Therefore the amount of taper per millimetre of unit length = $\frac{1}{k}$

The *total amount of taper* will be the taper per millimetre $\frac{1}{k}$ multiplied by the total length of taper (ℓ).

Total taper = $\frac{1}{k} \times \ell$ or $\frac{\ell}{k}$

$D = d + $ total amount of taper

$D = d + \frac{\ell}{k}$

EXAMPLE: Calculate the large diameter (D) for a 1:30 taper having a small diameter of 10 mm and a length of 60 mm.

SOLUTION:
Since taper is 1:30 · $k = 30$

$$D = d + \frac{\ell}{k}$$
$$= 10 + \frac{60}{30}$$
$$= 10 + 2$$
$$= 12 \text{ mm}$$

TAILSTOCK OFFSET CALCULATIONS

The tailstock offset method is often used to produce tapers in a lathe on work turned between centres when a taper attachment is not available. To produce a taper, the amount of tailstock offset must first be calculated by applying one of the following simple formulas.

INCH TAILSTOCK OFFSET CALCULATIONS

Tailstock offset = $\frac{\text{taper/ft.} \times \text{O.L.}}{12 \times 2}$

taper/ft. = taper per foot

O.L. = Overall length of work
12 = inches per foot
2 = the offset is taken from the centre line of the work

To calculate the tailstock offset for a 10 in. long piece of work which has a 3/4 in. taper/ft.:

Tailstock offset = $\frac{3/4 \times 10}{24}$
$= 3/4 \times 1/24 \times 10$
$= 5/16$ in.

In cases where it is not necessary to find the taper per foot, a simplified formula can be used to calculate the amount of tailstock offset:

Tailstock offset = $\frac{\text{O.L.}}{\text{T.L.}} \times \frac{(D-d)}{2}$

O.L. = overall length of work
T.L. = length of the tapered section
D = diameter at the large end of the taper
d = diameter at the small end of the taper

EXAMPLE: Using the simplified formula to find the tailstock offset for the following piece of work: large diameter is 1 in., small diameter is 23/32 in., the length of the taper is 6 in., and the overall length of the work is 18 in.

SOLUTION:

Tailstock offset = $\frac{18}{6} \times \frac{(1 - 23/32)}{2}$
$= \frac{18}{6} \times \frac{9}{64}$
$= 27/64$ in.

156 CHAPTER 11 / THE ENGINE LATHE

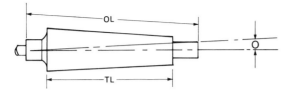

Fig. 11-78B
Metric taper turning by the tailstock offset method.

METRIC TAILSTOCK OFFSET CALCULATIONS

If the taper is to be turned by off-setting the tailstock, the amount of offset is calculated as follows (see Fig. 11-78B).

$$\text{Offset } (o) = \frac{D - d}{2 \times \ell} \times L$$

D = large diameter
d = small diameter
ℓ = length of taper
L = length of work

EXAMPLE: Calculate the tailstock offset required to turn a 1:30 taper 60 mm long on a workpiece 300 mm long. The small diameter of the tapered section is 20 mm.

SOLUTION:

Large diameter of taper $(D) = d + \dfrac{\ell}{k}$

$$= 20 + \frac{60}{30}$$
$$= 20 + 2$$
$$= 22 \text{ mm}$$

$$\text{Tailstock offset} = \frac{D - d}{2 \times \ell} \times L$$
$$= \frac{22 - 20}{2 \times 60} \times 300$$
$$= \frac{2}{120} \times 300$$
$$= 5 \text{ mm}$$

TAPER ATTACHMENTS

Turning a taper using the taper attachment provides many advantages in producing both internal and external tapers. The most important are:

1. Setup is simple. The taper attachment is easy to connect and disconnect.
2. Live and dead centres are not adjusted, so centre alignment is not disturbed.
3. Greater accuracy can be achieved, since one end of the guide bar is graduated in degrees and the other end in a ratio of 1 mm per unit of length, or in inches of taper per foot.
4. The taper can be produced between centres or on projecting work from any holding device, such as a chuck or a collet, regardless of the length of the work.
5. Internal tapers can be produced with the same taper setup as for external tapers.
6. A great range of tapers can be produced, and this is of special advantage when production is a factor and various tapers are required on a unit.

There are two common types of taper attachments in use:

(a) the plain taper attachment (Fig. 11-79);
(b) the telescopic taper attachment.

To use the plain taper attachment, the crossfeed screw nut must be disengaged from the cross-slide. When a telescopic taper attachment is used, the crossfeed screw is not disengaged, and the depth of cut can be set by the crossfeed handle.

Inch Taper Attachment Offset Calculations

Most tapers cut on a lathe with the taper attachment are expressed in taper per foot. If the taper per foot of the taper on the workpiece is not given, it may be calculated by using the following formula:

$$\text{Taper per foot} = \frac{(D - d) \times 12}{T.L.}$$

EXAMPLE: Calculate the taper per foot for a taper with the following dimensions: large diameter (D) 1-3/8 in., small diameter (d) 15/16 in., length of tapered section $(T.L.)$ 7 in.

$$\text{taper/ft.} = \frac{(1\text{-}3/8 - 15/16) \times 12}{7}$$
$$= \frac{7/16 \times 12}{7}$$
$$= 3/4 \text{ in.}$$

Metric Taper Attachment Offset Calculations

When the taper attachment is used to turn a taper, the amount the guide bar is set over may be determined as follows:

(a) If the angle of taper is given on the print, set the guide bar to one-half the included angle (Fig. 11-78C).

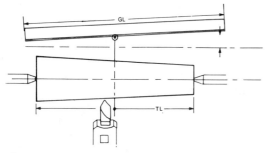

Fig. 11-78C

Metric taper turning by the taper attachment method.

(b) If the angle of taper is not given on the print, use the following formula to find the amount of guide bar setover.

$$\text{Guide bar setover} = \frac{D - d}{2} \times \frac{L}{\ell}$$

D = large diameter of taper
d = small diameter of taper
ℓ = length of taper
L = length of taper attachment guide bar

EXAMPLE: Calculate the amount of setover for a 500 mm long guide bar to turn a 1:50 × 250 mm long taper on a workpiece. The small diameter of the taper is 25 mm.

$$\text{Large diameter of taper} = d + \frac{\ell}{k}$$
$$= 25 + \frac{250}{50}$$
$$= 30 \text{ mm}$$

$$\text{Guide bar setover} = \frac{D - d}{2} \times \frac{L}{\ell}$$
$$= \frac{30 - 25}{2} \times \frac{500}{250}$$
$$= \frac{5}{2} \times 2$$
$$= 5 \text{ mm}$$

TAPER TURNING

Tapers can be cut on a lathe by using the taper attachment, offsetting the tailstock, and by setting the compound rest to the angle of the taper.

To Cut a Taper Using a Taper Attachment

The procedure for machining a taper using either a plain or telescopic taper attachment is basically the same with only minor adjustments required. The procedure for setting the plain taper attachment (Fig. 11-79) and cutting a taper are outlined as follows:

1. Clean and oil the *guide bar*.
2. Loosen the guide bar lock nuts so that it is free to move on the *base plate*.
3. By adjusting the *locking screws*, offset the end of the guide bar the required

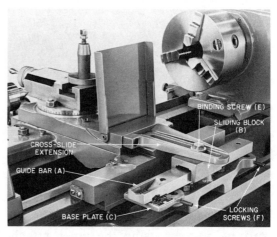

Courtesy Kostel Enterprises Ltd.

Fig. 11-79

Parts of a plain taper attachment.

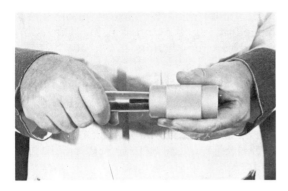

Courtesy Kostel Enterprises Ltd.

Fig. 11-80
Using a taper ring gauge to check the accuracy of a taper.

amount, or for inch tapers, set the taper attachment to the required taper per foot.
4. Tighten the guide bar lock nuts.
5. Swivel the compound rest so that it is at about 30° to the cross-slide.
6. Set the cutting tool to centre and tighten the toolpost securely.
7. Mount the work in the lathe and mark the length to be tapered.
8. Feed the cutting tool in until it is about 6 mm (1/4 in.) from the diameter of the work.
9. Remove the *binding screw*, which connects the cross-slide and the crossfeed screw nut.
10. Use the binding screw to connect the cross-slide extension to the *sliding block* using two-finger pressure on the wrench.
11. Insert a plug in the hole where the binding screw was removed to keep chips and dirt from damaging the crossfeed screw.
12. Move the carriage until the cutting tool clears the right-hand end of the work by about 12 mm (1/2 in.).
13. Take a light trial cut for about 1.5 mm (or 1/16 in.) and check the taper for size.
14. Set the depth of the roughing cut about 1.5 mm (or 1/16 in.) larger than the finish size, and rough cut the taper to the required length.
15. Check the taper for fit (Fig. 11-80). See the section "To Check a Taper With a Ring Gauge" in this chapter.
16. Readjust the taper attachment setting if necessary and take a light trial cut from the taper.
17. When the taper fit is correct, cut the taper to size.

TAILSTOCK OFFSET METHOD

The tailstock offset method of cutting tapers *should only be used* when a lathe is not equipped with a taper attachment and the work is mounted between centres. The tailstock centre must be moved out of line with the headstock centre enough to produce the desired taper (Fig. 11-81). Since the tailstock can only be offset a certain amount, the range of tapers which can be cut is limited.

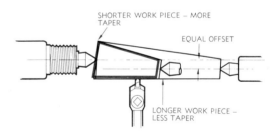

Fig. 11-81
The amount of taper cut using the tailstock offset method varies with the length of the workpiece.

To Offset the Tailstock
1. Calculate the amount the tailstock must be offset to cut the desired taper on the work.
2. Loosen the tailstock clamp nut.
3. Loosen one tailstock adjusting screw and tighten the opposite one until the tailstock offset is correct (Fig. 11-82).

TAPER TURNING 159

Courtesy Kostel Enterprises Ltd.

Fig. 11-82
Offsetting the tailstock for taper turning.

4. Tighten the adjusting screw that was loosened and recheck the offset with a rule.
5. Correct the setting if necessary, and then tighten the tailstock clamp nut.
6. Mount the work between centres and cut the taper to size by following operations 12 to 15 of the section "To Cut a Taper Using a Taper Attachment" in this chapter.

COMPOUND REST METHOD

The compound rest is used to cut short, steep tapers that are given in degrees. The compound rest must be set to the required angle, and then the cutting tool is advanced along the taper using the compound rest feed handle.

To Cut a Taper Using the Compound Rest

1. Check the print for the angle of the taper in degrees.
2. Loosen the compound rest lock nuts.
3. Swivel the compound rest to the required angle (Fig. 11-83).
   (a) If the included angle is given as in Fig. 11-83A, set the compound rest to one-half the included angle.
   (b) If the angle is given on one side only, as in Fig. 11-83B, set the compound rest to that angle.
4. Tighten the compound rest lock nuts, using only a two-finger pull on the wrench to avoid stripping the thread on the compound rest studs (Fig. 11-84).

Courtesy Kostel Enterprises Ltd.

Fig. 11-84
When tightening the compound rest locknuts, use only a two-finger pull on the wrench.

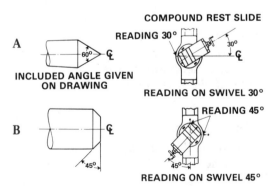

Fig. 11-83
The direction to swing the compound rest for cutting various angles.

5. Set the toolbit on centre and then swivel the toolholder so that is it at 90° to the compound rest (Fig. 11-85).
6. Bring the toolbit close to the diameter to be cut, using the carriage handwheel and crossfeed handle.
7. Cut the taper by turning the *compound rest feed screw*.
8. Check the taper for size and angle.

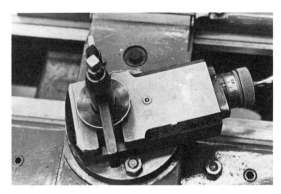

Courtesy Kostel Enterprises Ltd.

Fig. 11-85
Set the toolholder at 90° to the side of the compound rest and the cutting tool on centre.

CHECKING A TAPER

External tapers can be checked for accuracy of size or fit by using a taper ring gauge, a standard micrometer, or a special taper micrometer.

To Check a Taper With a Ring Gauge

1. Draw three equally spaced light lines with chalk or mechanics' blue along the length of the taper (Fig. 11-86).
2. Insert the taper into the gauge and turn counterclockwise one-half turn, then remove it for inspection.

Courtesy Kostel Enterprises Ltd.

Fig. 11-86
Checking a taper for accuracy using a taper ring gauge.

3. If the chalk is rubbed from the whole length of the taper, the taper is correct.
4. If the chalk lines are rubbed from only one end, the taper is incorrect.
5. By making slight adjustments to the taper setup and taking trial cuts, machine the taper until the fit is correct.

To Check an Inch Taper With a Standard Micrometer

1. Calculate the amount of taper per inch of the taper.
2. Clean the tapered section of the work and apply layout dye as shown in Fig. 11-87.

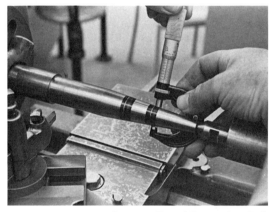

Courtesy Kostel Enterprises Ltd.

Fig. 11-87
Checking the accuracy of a taper using a micrometer.

3. Lay out two lines exactly 1 in. apart (Fig. 11-87).
4. Measure the taper with a micrometer at both lines so that the left edge of the micrometer anvil and spindle just touch the line.
5. Subtract the difference between the two readings, and compare the answer with the required taper per inch.
6. If necessary, adjust the taper attachment setting to correct the taper.

Greater accuracy is possible if the length of the work permits the lines to be laid out 2 or 3 in. apart. Determine the difference in diameters at the lines and divide by the distance between the lines (in inches) to determine the taper per inch.

To Check a Metric Taper With a Metric Micrometer

1. Check the drawing for the taper required.
2. Clean the tapered section of the work and apply layout dye.
3. Lay out two lines on the taper which are the same distance apart as the second number in the taper ratio.
   EXAMPLE: If the taper was 1:20, the lines would be 20 mm apart.

   If the work is long enough, lay out the lines at double or triple the length of the tapered section and increase the difference in diameters by the appropriate amount. For instance, on a 1:20 taper the lines may be laid out 60 mm apart or three times the unit length of the taper. Therefore the difference in diameters would then be 3 × 1, or 3 mm. This will give a more accurate check of the taper.
4. Measure the diameters carefully with a metric micrometer at the two lines.
   The difference between these two diameters should be 1 mm for each unit of length.
5. If necessary, adjust the taper attachment setting to correct the taper.

MOUNTING AND REMOVING CHUCKS

Lathe accessories such as chucks and drive plates are fitted to the headstock of a lathe. The proper procedure for mounting and removing these accessories must be followed in order not to damage the lathe spindle and/or accessories, and to preserve the accuracy of the machine. There are three types of lathe spindle noses: the threaded spindle nose, the tapered spindle nose, and the cam-lock spindle nose.

Courtesy Kostel Enterprises Ltd.

Fig. 11-88
A chuck cradle prevents injury to the hands and damage to the lathe bed when mounting or removing a chuck.

To Remove a Chuck

1. Set the lathe in the slowest speed. SHUT OFF THE ELECTRICAL SWITCH.
2. Place a chuck cradle under the chuck (Fig. 11-88).
3. Remove the chuck or accessory by following the steps outlined below, depending on the type of lathe spindle nose.

Courtesy South Bend Lathe, Inc.

Fig. 11-89
A hardwood block can be used to remove a chuck from a threaded spindle.

THREADED SPINDLE NOSE
(a) Turn the lathe spindle until a chuck-wrench socket is in the top position.
(b) Insert the chuck wrench into the hole and pull it *sharply counterclockwise* (toward you).

OR

(a) Place a block or short stick under the chuck jaw as shown in Fig. 11-89.
(b) Revolve the lathe spindle by hand in a *clockwise* direction until the chuck is loosened on the spindle.
(c) Remove the chuck from the spindle and store it where it will not be damaged.

TAPER SPINDLE NOSE
(a) Secure the proper C-spanner wrench.
(b) Place it around the front of the lock ring of the spindle with the handle in an upright position (Fig. 11-90).
(c) Place one hand on the curve of the spanner wrench to prevent it from slipping off the lock ring.
(d) With the palm of the other hand, *sharply* strike the handle of the wrench in a *clockwise* direction.
(e) Hold the chuck with one hand, while turning the lock ring clockwise with the other hand.
(f) If the lock ring becomes tight, use the spanner wrench to break the taper contact between the spindle and chuck.
(g) Store the chuck with the jaws in the up position.

CAM-LOCK SPINDLE NOSE
(a) With the proper size wrench turn each cam-lock *counterclockwise* until its registration line matches the registration line on the spindle nose, or is at the 12 o'clock position.
(b) Hold the chuck and with the other hand *sharply* strike the chuck to remove it from the spindle (Fig. 11-91).
(c) Remove and store the chuck properly.

Courtesy Kostel Enterprises Ltd.

Fig. 11-90
A C-spanner wrench is used to loosen the lock ring on a taper spindle nose.

Courtesy Kostel Enterprises Ltd.

Fig. 11-91
A sharp blow with the hand is used to break the taper contact between the lathe spindle and the chuck.

MOUNTING AND REMOVING CHUCKS

To Mount a Chuck

The following procedures apply to any accessories which are mounted *on* a lathe spindle nose.

1. Set the lathe to the slowest speed. SHUT OFF THE ELECTRICAL SWITCH.
2. Clean all surfaces of the spindle nose and the mating parts of the chuck.
3. Place a cradle block on the lathe bed in front of the spindle and place the chuck on the cradle (Fig. 11-88).
4. Slide the cradle close to the lathe spindle nose and mount the chuck.
5. Mount the chuck or accessory by following the steps outlined below, depending on the type of lathe spindle nose.

THREADED SPINDLE NOSE

(a) Revolve the lathe spindle *by hand* in a counterclockwise direction and bring the chuck up to the spindle. *NEVER USE POWER.*
(b) If the chuck and spindle are clean and correctly aligned, the chuck should easily thread onto the lathe spindle.

Courtesy Kostel Enterprises Ltd.

Fig. 11-93
Tightening the lock ring by hand.

(c) When the chuck adaptor plate is within 1.5 mm (1/16 in.) of the spindle shoulder, give the chuck a quick turn to seat it against the spindle shoulder.
(d) Do not jam a chuck against the shoulder too tightly. It may damage the threads and make the chuck difficult to remove.

TAPER SPINDLE NOSE

(a) Revolve the lathe spindle by hand until the key on the spindle nose aligns with the keyway in the tapered hole of the chuck (Fig. 11-92).

Courtesy Kostel Enterprises Ltd.

Fig. 11-92
The spindle key and keyway must be aligned when mounting a tapered spindle nose chuck.

Courtesy Kostel Enterprises Ltd.

Fig. 11-94
Tightening the lock ring securely with a C-spanner.

(b) Slide the chuck onto the lathe spindle, and at the same time turn the lock ring in a *counterclockwise* direction (Fig. 11-93).
(c) Tighten the lock ring securely with a spanner wrench by striking it sharply downward (Fig. 11-94) when standing at the front of the machine.

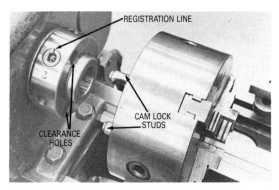

Courtesy Kostel Enterprises Ltd.

Fig. 11-95
Aligning the lathe spindle holes and the cam lock studs of the chuck.

CAM-LOCK SPINDLE NOSE
(a) Align the registration of each cam lock with the registration line on the lathe spindle nose (Fig. 11-95).
(b) Revolve the lathe spindle by hand until the clearance holes in the spindle align with the cam lock studs of the chuck (Fig. 11-95).
(c) Slide the chuck onto the spindle.
(d) Securely tighten each cam lock in a clockwise direction (Fig. 11-96).

THREE-JAW CHUCK WORK

The three-jaw universal chuck is used to hold round or hexagonal work for machining. With proper care, this chuck should be able to hold work to within 0.05 mm (0.002 in.) of concentricity even after long use. Three-jaw chucks are supplied with two sets of jaws: a regular set and a reversed set. The *regular set* is used to grip outside diameters and also inside diameters of large work. The *reversed set* is used to grip the outside of large-diameter work. All chuck jaws are stamped with the same serial number as the chuck and also numbered 1, 2, or 3, to match the slot in which they fit. They have been fitted and ground true for that chuck and must *never* be used on another chuck.

Courtesy Kostel Enterprises Ltd.

Fig. 11-96
To tighten the chuck on the spindle, turn the cam locks in a clockwise direction.

To Face Work In a Chuck
The purpose of facing work in a chuck is to obtain a true flat surface and cut the work to length.

Courtesy Kostel Enterprises Ltd.

Fig. 11-97
The work should extend no more than three times its diameter beyond the chuck jaws.

1. Set the work in the chuck so that no more than three times its diameter extends beyond the chuck jaws. (Distance X in Fig. 11-97).
2. Swivel the compound rest 30° to the right if only one surface on the work must be faced.

OR

Swivel the compound rest 90° to the cross-slide if a series of steps or shoulders must be faced to accurate length on the same workpiece.
3. Fasten a facing toolbit in the toolholder and set its point to centre height.
4. Adjust the toolholder until the point of the facing tool is closest to the work and there is a space left along the side (Fig. 11-98).

Courtesy Kostel Enterprises Ltd.

Fig. 11-98
A toolbit set up for facing the end of a workpiece held in a chuck.

5. Move the carriage until the toolbit starts a light cut at the centre of the surface to be faced.
6. Lock the carriage in position and set the depth of cut with the compound rest handle:
   (a) Twice the amount to be removed if the compound rest is set at 30°.
   (b) The same as the amount to be removed if the compound rest is set at 90° to the cross-slide.
7. Face the work to length.

To Drill Centre Holes

Centre holes can be drilled on a lathe in round or hexagonal work without having to lay out the location of the centre.
1. Grip the work short in a three-jaw chuck.
   No more than three times the diameter should extend beyond the chuck jaws.
2. Square the end of the work by facing (Fig. 11-99).
3. Mount a drill chuck in the tailstock spindle.

Courtesy Kostel Enterprises Ltd.

Fig. 11-99
Before drilling a centre hole, the work must be faced.

Courtesy Kostel Enterprises Ltd.

Fig. 11-100
The centre drill is mounted on a drill chuck in the tailstock.

4. Select the proper centre drill to suit the work diameter and fasten it in the drill chuck (Fig. 11-100).

   See Table 10-3 in Chapter 10, *Drill Presses*.
5. Check the lines on the back of the tailstock to see that they are aligned (Fig. 11-101). Correct if necessary.

Courtesy Kostel Enterprises Ltd.

Fig. 11-101
The upper and lower lines on the tailstock should be aligned before centre drilling.

6. Set the lathe speed to approximately 1200 to 1500 r/min.
7. Move the tailstock until the centre drill is close to the work and then lock the tailstock clamp nut.
8. Start the lathe spindle and turn the tailstock handwheel to feed the centre drill into the work (Fig. 11-102).
9. Frequently apply cutting fluid, and drill the centre hole until the top of the hole is to the correct diameter (about 5 mm (3/16 in.) for 19 mm (3/4 in.) diameter work).

To Drill a Hole

A hole may be drilled quickly and accurately in work held in a chuck. Straight shank drills are generally held in a drill chuck mounted in the tailstock, while taper shank drills are mounted directly in the tailstock spindle.

Courtesy Kostel Enterprises Ltd.

Fig. 11-102
Drilling the centre hole in the workpiece.

1. Mount the work true in a chuck.
2. Face the end of the workpiece.
3. Set the lathe to the proper speed for the type of material to be drilled.
4. Check the tailstock centre and make sure that it is in line.
5. With a centre drill, spot the hole until about one-half of the tapered portion of the centre drill enters the work.
6. (a) If the hole to be drilled is about 12 mm (1/2 in.) or less, mount the correct size of drill in a drill chuck mounted in the tailstock spindle and support it with the back end of a toolholder (Fig. 11-103).

Courtesy Kostel Enterprises Ltd.

Fig. 11-103
A drill mounted in the drill chuck is supported by the back of the toolholder when starting to drill.

(b) If a hole over 12 mm (1/2 in.) is to be drilled, mount the tapered shank of the drill in the tailstock spindle and fasten a lathe dog on the body close to the tailstock spindle. The tail of the dog should rest on the top of the compound rest (Fig. 11-104). This setup will prevent the drill from turning and damaging the taper in the tailstock spindle.

Courtesy Kostel Enterprises Ltd.

Fig. 11-104
A taper shank drill mounted in a tailstock can be prevented from turning with a lathe dog.

7. Start the lathe and turn the tailstock handwheel to feed the drill into the work.
8. Apply cutting fluid frequently and drill the hole to depth.
9. Check the depth of the hole using a rule or the graduations on the tailstock spindle.
10. Always *ease up* the drill pressure as a drill starts to break through the work.

To Ream a Hole
Reaming may be performed on a lathe to bring a drilled or bored hole to an accurate size and to produce a good surface finish.
1. *Check that the tailstock centre is in line.*
2. Mount the work in a chuck.
3. Face and centre drill the work.

4. Select the proper size drill to leave material in a hole for reaming:
   (a) 0.40 mm smaller than the finish size for holes up to 12 mm diameter (or 1/64 in. smaller for holes up to 1/2 in. diameter).
   (b) 0.80 mm smaller than the finish size for holes over 12 mm diameter (or 1/32 in. smaller for holes over 1/2 in. diameter).
5. Apply cutting fluid and drill the hole to the proper depth.
6. Mount the reamer in the drill chuck or tailstock spindle (Fig. 11-105).
7. Set the lathe to one-half the drilling speed.
8. Apply cutting fluid and turn the tailstock handwheel to feed the reamer into the hole.
9. Remove the reamer and store it where it will not be nicked or damaged.

Courtesy Kostel Enterprises Ltd.

Fig. 11-105
Reaming brings a hole to size and produces a good finish.

Turning Work Held in a Chuck
Much of the work machined on a lathe is held in some type of chuck. The procedure for turning a diameter on work held in a chuck is the same as that used for turning between lathe centres. However, the workpiece should not extend beyond the chuck

jaws more than three times the diameter of the work, to prevent the work from springing or bending. If the amount of work extending beyond the chuck jaws is long, support the right-hand end of the work with a revolving tailstock centre or a steady rest. Wherever possible, the work should be machined to size in two cuts; one rough and one finish cut.

To Cut Off Work Held in a Chuck

Cut-off tools, often called parting tools, are used for cutting off work projecting from a chuck, for grooving, and for undercutting. The inserted blade type of parting tool is most commonly used; it is provided in three types (Fig. 11-106).

Courtesy Kostel Enterprises Ltd.

Fig. 11-107
The cut-off tool should be gripped short to reduce vibration.

Courtesy J.H. Williams & Co.

Fig. 11-106
Types of inserted-blade cut-off tools.

1. Mount the work in the chuck, with the part to be cut off as close to the jaws as possible.
2. Mount the cut-off tool on the left-hand side of the compound rest and as close to the toolpost as possible, to minimize vibration (Fig. 11-107).
3. Have the cutting blade extending beyond the holder half the diameter of the work to be cut, plus 3 mm (1/8 in.) for clearance (Fig. 11-108).
4. Set the cutting tool to centre and at 90° to the centre line of the work, and tighten the toolpost screw securely.

Courtesy Kostel Enterprises Ltd.

Fig. 11-108
The blade should extend only slightly more than half the diameter of the workpiece.

Fig. 11-109
Using a rule to position the cut-off tool.

5. Set the lathe to one-half of the turning speed.
6. Move the cutting tool into position for the proper length of cut (Fig. 11-109).
7. Lock the carriage by tightening the carriage lock screw.
8. Start the lathe, and feed the cut-off tool steadily into the work using the cross-feed handle. Cut brass and cast iron dry, but use cutting fluid for steel.
9. Before the cut is completed, remove the burrs from each side of the groove with a file.

To avoid chatter, keep the tool cutting steadily and apply cutting fluid during the operation. Feed slowly when the part is almost cut off.

To Tap a Hole In a Lathe

Internal threads may be cut on a lathe by using the proper size tap. A standard tap may be used for this operation; however, a gun tap is preferred because the chips are cleared ahead of the tap. The tap is aligned by placing the point of the dead centre in the shank end of the tap to guide it while the tap is turned by means of a tap wrench. The lathe spindle is locked, and the tap is turned by hand (Fig. 11-110).

1. Mount the work in the chuck and face the end.
2. Centre drill a hole so that the top edge of the hole is slightly larger than the tap diameter.
3. Select the proper tap drill size for the tap to be used.
   See the Appendix, Table 4.
4. Set the lathe to the proper speed for the diameter of the drill being used.
5. Drill the tap drill hole to the required depth, using cutting fluid if required.
6. Stop the lathe and lock the spindle, or put the lathe in its slowest speed.
7. Select the proper size taper tap and mount it in a tap wrench.
8. Place the tapered end of the tap in the hole of the work and support the other end with the dead centre of the lathe (Fig. 11-111).
9. Apply cutting fluid and start the tap into the hole by slowly turning the tap wrench clockwise.
10. Keep the dead centre in light contact with the shank of the tap during this

Fig. 11-110
Cutting an internal thread on a lathe with a tap.

operation by turning the tailstock handwheel while turning the tap with the other hand (Fig. 11-112).
11. Back off the tap every half turn to break the chips and apply cutting fluid frequently.
12. Remove the taper tap and complete tapping the hole with a plug or bottoming tap.

Fig. 11-111
The end of the tap is supported by the dead centre of the lathe.

Fig. 11-112
The dead centre is kept snug in the shank of the tap by the tailstock handwheel.

THREADS AND THREAD CUTTING

STANDARD THREAD FORMS

A thread is a helical ridge of uniform section formed on the inside or outside of a cylinder or a cone. Some of the common types of thread forms are shown in Fig. 11-113.

(a) *American National Thread* (Fig. 11-113A) is listed under three main divisions: National Coarse, National Fine, National Series. This thread is commonly known as a locking thread form in America. The new ISO metric threads will be used for the same purposes as these threads.

(b) *Unified Screw Thread* (Fig. 11-113B) was the result of a need for a common system for use in Canada, United States, and England. This thread incorporates the features of the American National Form and the British Standard Whitworth threads. Threads in the Unified series are interchangeable with American National and Whitworth threads of the same pitch and diameter.

(c) *International Metric Thread* (Fig. 11-113C) is a standard thread currently used throughout Europe. It is used in North America mainly on instruments and spark plugs.

(d) *American National Acme Thread* (Fig. 11-113D) is generally classified as a power transmission type.

(e) *Square Thread* (Fig. 11-113E) is used for maximum transmission power. Because of its shape, friction between its matching threads is kept to a minimum.

(f) *ISO Metric Threads* (Fig. 11-113F). Over the past several decades, one of the world's major industrial problems has been the lack of an international thread standard whereby the thread standard used in any country could be interchanged with that of another country. In April 1975, the Industrial Organization for Standardization (ISO) drew up an agreement covering a standard metric thread profile, the sizes and pitches for the various threads in

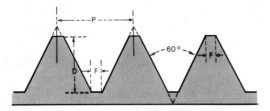
A - American national form thread

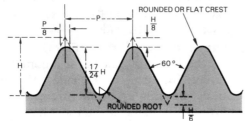
B - Unified screw thread

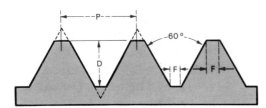
C - International metric thread

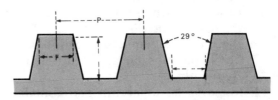
D - American national acme thread

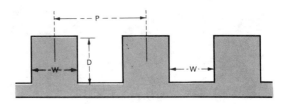
E - Square thread

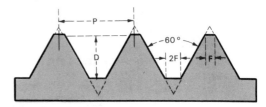
F - ISO metric thread

Fig. 11-113
Common thread forms and dimensions.

the new ISO Metric Thread Standard. The new series has only twenty-five thread sizes ranging in diameter from 1.6 mm to 100 mm. See Table 11-4 for this series.

These metric threads are identified by the letter M, the nominal diameter and the pitch. For example, a metric thread with an outside diameter of 5 mm and a pitch of 0.8 mm would be identified as follows: M 5 × 0.8.

The new ISO series will not only simplify thread design but will generally produce stronger threads for a given diameter and pitch and will reduce the large inventory of fasteners now required by industry.

The new ISO metric thread (Fig. 11-113F) has a 60° included angle and a crest equal to 0.125 times the pitch, which is similar to the National Form thread. The main difference, however, is the depth of thread, which is 0.54127 times the pitch. Because of these dimensions, the root of the thread is larger than that of the National Form thread. The root of the new ISO metric thread is 1/4 of the pitch (0.250P).

THREAD TERMS AND CALCULATIONS

Screw threads form a very important part of every component made, from a tiny wristwatch to a large earthmover. To understand thread theory and screw cutting, you must know the parts of a thread (Fig. 11-114). All threads have common thread terms. The American National Thread Series and the new ISO metric thread are the only threads fully explained in this book.

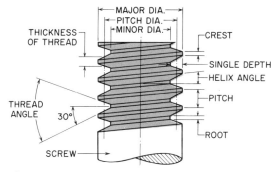

Fig. 11-114
The main parts of a screw thread.

ANGLE OF THREAD — The angle included between the sides of the thread; for example, the thread angle of the new ISO Metric Thread and that of the American National Form is 60°.

MAJOR DIAMETER — The largest diameter of the thread on the screw or nut.

MINOR DIAMETER — The smallest diameter of an external or internal screw thread.

NUMBER OF THREADS — The number of roots or crests per inch of the threaded length. This term does not apply to metric threads.

PITCH — The distance from a point on one thread to the corresponding point on the next thread measured parallel to the axis. It is expressed in millimetres for metric threads.

LEAD — The distance a screw thread advances axially for one complete revolution.

CREST — The top surface joining the two sides of a thread.

ROOT — The bottom surface joining the sides of two adjacent threads.

SIDE — The surface of the thread which connects the crest with the root.

DEPTH OF THREAD — The distance between the crest and the root of a thread, measured perpendicular to the axis.

**TABLE 11-4
ISO METRIC PITCH-DIAMETER COMBINATIONS**

Nominal Diameter	Thread Pitch	Nominal Diameter	Thread Pitch
1.6	0.35	20	2.5
2.0	0.40	24	3.0
2.5	0.45	30	3.5
3.0	0.50	36	4.0
3.5	0.60	42	4.5
4.0	0.70	48	5.0
5.0	0.80	56	5.5
6.0	1.00	64	6.0
8.0	1.25	72	6.0
10.0	1.50	80	6.0
12.0	1.75	90	6.0
14.0	2.00	100	6.0
16.0	2.00		

Calculations for American National Form Thread

P = Pitch of Thread
$= \dfrac{1}{\text{Number of threads per inch}}$

D = Depth of Thread
$= 0.6495 \times$ Pitch
$= \dfrac{0.6495}{\text{Number of threads per inch}}$

F = Width of Flat on Crest or Root
$= \dfrac{P}{8}$
$= \dfrac{1}{8 \times \text{Number of threads per inch}}$

N = Number of threads per inch

EXAMPLE: Find the pitch, depth, and minor diameter of a 1 in. — 8 N.C. thread.

Major Diameter = 1.000 in.
$$\text{Pitch} = \frac{1}{N}$$
$$= \frac{1}{8} \text{in.}$$

THREADS AND THREAD CUTTING

Depth of Thread $= 0.6495 \times P$
$= 0.6495 \times \dfrac{1}{8}$
$= 0.081$ in.

Minor Diameter $=$ Major Dia. $- (D + D)$
$= 1.000 - (0.081 + 0.081)$
$= 0.838$ in.

Calculations for ISO Metric Threads

$P =$ Pitch of thread in millimetres
$D =$ Depth of thread
$\quad = 0.54127 \times$ pitch
$FC =$ Width of flat at the crest
$\quad = 0.125 \times$ pitch
$FR =$ Width of flat at the root
$\quad = 0.250 \times$ pitch

EXAMPLE: What is the pitch, depth, minor diameter, width of crest, and width of root for a M 14 × 2 thread?

Pitch $= 2$ mm
Depth $= 0.54127 \times 2$
$\quad = 1.082$ mm
Minor diameter $=$ Major diameter $- (D + D)$
$\quad = 14 - (1.082 + 1.082)$
$\quad = 11.84$ mm
Width of crest $= 0.125 \times$ pitch
$\quad = 0.125 \times 2$
$\quad = 0.25$ mm
Width of root $= 0.250 \times$ pitch
$\quad = 0.250 \times 2$
$\quad = 0.5$ mm

To Set the Quick-Change Gearbox for Threading

The quick-change gearbox (Fig. 11-115) is designed to speed up changing gears for thread cutting. This unit has gear changes which transmit a direct motion or a ratio from the spindle to the lead screw.
1. Check the print for the pitch in millimetres or number of threads per inch required.
2. On the quick-change gearbox chart find

Courtesy A.R. Williams Machinery Company

Fig. 11-115
The quick-change gear box establishes the proper ratio between the spindle and the lead screw to cut a given thread.

the *whole number* which represents the pitch in millimetres or number of threads per inch.
3. Engage the tumbler lever in the hole at the bottom of the vertical column in which the number is located (Fig. 11-115).
4. Set the top lever in the proper position as indicated on the chart for the thread required.
5. Engage sliding gear in or out as required.
6. Turn the drive plate or chuck by hand, and make sure that the lead screw revolves.
7. Recheck the complete setup before thread cutting.

THREAD-CHASING DIAL

The thread-chasing dial is an indicator with a revolving dial, which can be either fastened to the carriage or built into it (Fig. 11-116). The chasing dial shows the operator when to engage the split-nut lever in order to take successive cuts in the same groove or thread. It also indicates the relationship between the ratio of the number of turns of the work and the lead screw with respect to the position of the cutting tool and the thread groove.

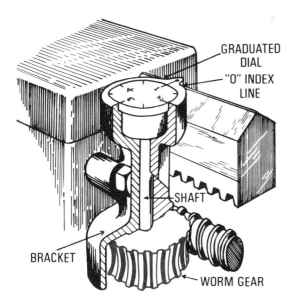

Fig. 11-116
The thread-chasing dial mechanism.

The thread-chasing dial is connected to a worm gear, which meshes with the threads of the lead screw. The dial is graduated into eight divisions, four numbered and four unnumbered, and it revolves as the lead screw turns. Figure 11-121 indicates when the split-nut lever should be engaged for cutting various numbers of threads per inch.

THREAD CUTTING

Thread cutting on a lathe is a process of producing a helical ridge of uniform section by cutting a continuous groove around a cylinder. This is done by taking successive light cuts with a threading toolbit the same shape as the thread form. In order to produce an accurate thread, it is important that the lathe, the cutting tool, and the work be set up properly.

To Set Up a Lathe for Threading (60° Thread)

1. Set the lathe speed to about 1/4 of the speed used for turning.
2. Set the quick-change gearbox for the required pitch in millimetres or number of threads per inch.
3. Engage the lead screw.
4. Secure a 60° threading toolbit, check the angle using a thread centre gauge, and mount it in a left-hand offset toolholder.
5. Set the compound rest at 29° to the right (to the left for a left-hand thread).
6. Mount the toolholder in the toolpost and set the point of the toolbit even with the dead centre point.
7. Set the toolbit at right angles to the centre line of the work, using a thread centre gauge (Fig. 11-117).

   Never jam a toolbit into a thread centre gauge. This can be avoided by aligning only the cutting (leading) side of the toolbit with the gauge. A piece of paper on the cross-slide under the gauge and toolbit makes it easier to check the tool alignment.
8. Arrange the apron feed lever in the neutral position and check the engagement of the split-nut lever.

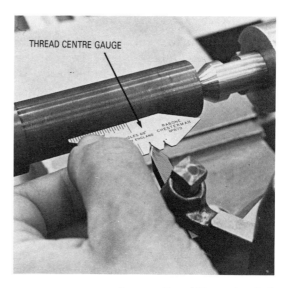

Courtesy Kostel Enterprises Ltd.

Fig. 11-117
Setting the threading tool square to the work with a centre gauge.

Fig. 11-118
Mark the driveplate slot, into which the lathe dog fits, with chalk.

To Cut a Thread

Thread cutting is a lathe operation that requires a great deal of attention and skill. It involves manipulation of the lathe parts, correlation of the hands, and strict attention to the operation being performed. Before proceeding to cut the thread, it is wise to take several trial passes without cutting, in order to get the feel of the machine.

1. Mount the work in the lathe and check that the diameter to be threaded is 0.05 mm (or 0.002 in.) undersize.

Fig. 11-119
Chamfering the end of the work with the side of the threading tool.

2. With chalk, mark the drive plate slot that is driving the lathe dog (Fig. 11-118).
3. Mark the length to be threaded by cutting a light groove at this point with the threading tool while the lathe is revolving (Fig. 11-118).
4. Chamfer the end of the work with the side of the threading tool (Fig. 11-119).
5. Move the carriage until the point of the threading tool is near the right-hand end of the work.
6. Turn the *crossfeed handle* until the threading tool is close to the diameter, but stop when the handle is at the 3 o'clock position (Fig. 11-120).

Fig. 11-120
Thread cutting is made easier when the crossfeed handle is at the 3 o'clock position.

7. Hold the crossfeed handle in this position and set the graduated collar to zero.
8. Turn the compound rest handle until the threading tool *lightly marks the work*, and set the compound rest graduated collar to zero.
9. Move the carriage to the right until the toolbit clears the end of the work.

10. Feed the compound rest *clockwise* about 0.07 mm (or 0.003 in.).
11. Engage the split-nut lever on the correct line of the thread-chasing dial (Fig. 11-121), and take a trial cut along the length to be threaded.
12. At the end of the cut, turn the crossfeed handle *counterclockwise* to move the toolbit away from the work and disengage the split-nut lever (Fig. 11-122).
13. Stop the lathe and check the number of threads per inch with a thread pitch gauge, rule, or centre gauge (Fig. 11-123).

If the pitch in millimetres (or number of threads per inch) produced by the trial cut is not correct, recheck the quick-change gearbox setting.

14. After each cut, turn the carriage handwheel to bring the toolbit to the start of the thread and return the crossfeed handle to zero.

THREADS PER INCH TO BE CUT	WHEN TO ENGAGE SPLIT NUT		READING ON DIAL
EVEN NUMBER OF THREADS	ENGAGE AT ANY GRADUATION ON THE DIAL	1 1 ½ 2 2 ½ 3 3 ½ 4 4 ½	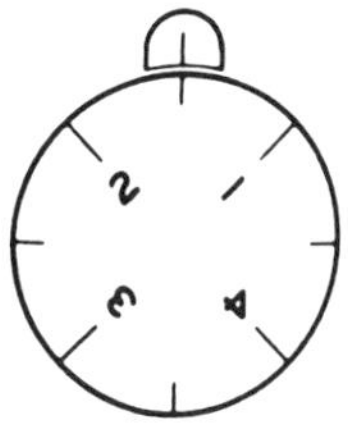
ODD NUMBER OF THREADS	ENGAGE AT ANY MAIN DIVISION	1 2 3 4	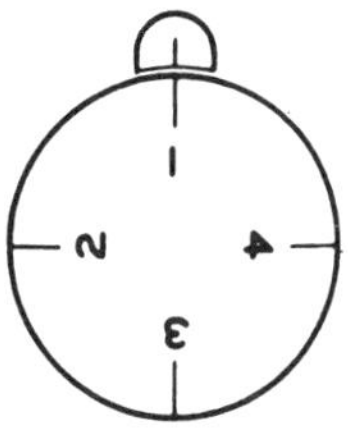
FRACTIONAL NUMBER OF THREADS	1/2 THREADS, E.G., 11 1/2 ENGAGE AT EVERY OTHER MAIN DIVISION 1 & 3, OR 2 & 4 OTHER FRACTIONAL THREADS ENGAGE AT SAME DIVISION EVERY TIME		
THREADS WHICH ARE A MULTIPLE OF THE NUMBER OF THREADS PER INCH IN THE LEAD SCREW	ENGAGE AT ANY TIME THAT SPLIT NUT MESHES		USE OF DIAL UNNECESSARY

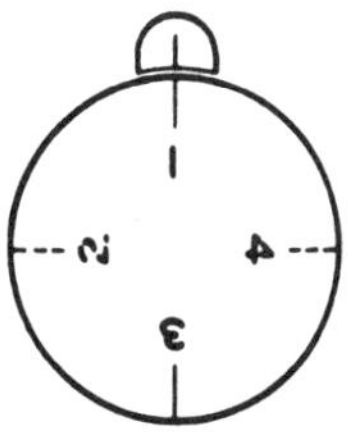

Fig. 11-121
Split nut engagement rules for inch thread cutting.

THREAD CUTTING

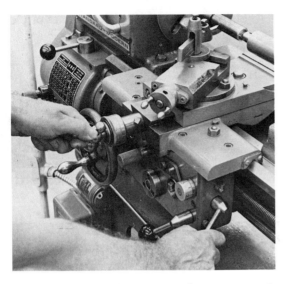

Courtesy Kostel Enterprises Ltd.

Fig. 11-122
Withdraw the toolbit; then disengage the split nut lever at the end of the section to be threaded.

15. Set the depth of all threading cuts with the compound rest handle.
    For National Form threads, use Table 11-5A; for ISO metric threads, see Table 11-5B.
16. Apply cutting fluid and take successive cuts until the top (crest) and the bottom (root) of the thread are the same width.

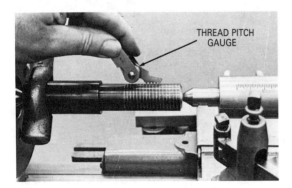

Fig. 11-123
Checking the number of threads per inch with a thread pitch gauge.

17. Remove the burrs from the top of the thread with a file (Fig. 11-124).
18. Check the thread with a master nut and take further cuts, if necessary, until the nut fits the thread freely with no end play (Fig. 11-125).

Fig. 11-124
Removing the burrs from the top of the thread with a fine file.

TABLE 11-5A DEPTH SETTINGS WHEN CUTTING 60° NATIONAL FORM THREADS			
	Compound Rest Setting		
tpi	0°	30°	29°
24	0.027	0.031	0.031
20	0.033	0.038	0.037
18	0.036	0.042	0.041
16	0.041	0.047	0.046
14	0.047	0.054	0.053
13	0.050	0.058	0.057
11	0.059	0.068	0.067
10	0.065	0.075	0.074
9	0.072	0.083	0.082
8	0.081	0.094	0.092
7	0.093	0.107	0.106
6	0.108	0.125	0.124
4	0.163	0.188	0.186

When using this table for cutting National form threads, the correct width of flat (0.125P) must be ground on the toolbit point; otherwise the thread will not be the correct width.

Fig. 11-125
Testing the thread with a master nut.

To Convert an Inch-Designed Lathe to Metric Threading

Metric threads may be cut on a standard quick-change gear lathe by using a pair of change gears having 50 and 127 teeth respectively. Since the lead screw has inch dimensions and is designed to cut threads per inch, it is necessary to convert the pitch in millimetres into threads per inch. To do this, it is first necessary to understand the relationship of the inch and the metric systems of measurement. 1 in. = 2.54 cm.

Therefore the ratio of inches to centimetres is 1:2.54, or $\frac{1}{2.54}$.

To cut a metric thread on a lathe, it is necessary to incorporate certain gears in the gear train which will produce a ratio of 1/2.54. These gears are:

$$\frac{1}{2.54} \times \frac{50}{50} = \frac{50 \text{ teeth}}{127 \text{ teeth}}$$

In order to cut metric threads, two gears having 50 and 127 teeth must be placed in the gear train of the lathe. The 50-tooth gear is used as the spindle or drive gear, and the 127-tooth gear is placed on the lead screw.

To Cut a 2.5 mm Metric Thread on a Standard Quick-Change Gear Lathe

1. Mount the 127-tooth gear on the lead screw (Fig. 11-126).
2. Mount the 50-tooth gear on the spindle (Fig. 11-126).
3. Convert the 2.5 mm pitch to threads per centimetre.
   10 mm = 1 cm
   Pitch = $\frac{10}{2.5}$ = 4 threads/cm
4. Set the quick-change gearbox to 4 threads/in. By means of the 50- and 127-tooth gears, the lathe will now cut 4 threads/cm, or 2.5 mm pitch.
5. Set up the lathe for thread cutting. See the section entitled "To Set Up a Lathe for Threading (60° Thread)".

TABLE 11-5B
DEPTH SETTINGS WHEN CUTTING 60° ISO METRIC THREADS

Pitch (mm)	Compound Rest Setting (mm)		
	0°	30°	29°
0.35	0.19	0.21	0.21
0.4	0.21	0.25	0.24
0.45	0.24	0.28	0.27
0.5	0.27	0.31	0.30
0.6	0.32	0.37	0.36
0.7	0.37	0.43	0.42
0.8	0.43	0.50	0.49
1.0	0.54	0.62	0.62
1.25	0.67	0.78	0.77
1.5	0.81	0.93	0.92
1.75	0.94	1.09	1.08
2.0	1.08	1.25	1.24
2.5	1.35	1.56	1.55
3.0	1.62	1.87	1.85
3.5	1.89	2.19	2.16
4.0	2.16	2.50	2.47
4.5	2.44	2.81	2.78
5.0	2.71	3.13	3.09
5.5	2.98	3.44	3.40
6.0	3.25	3.75	3.71

Fig. 11-126
The gears required to convert an inch lathe to cut metric threads.

6. Take a light trial cut. At the end of the cut, back out the cutting tool and stop the machine but *do not disengage the split nut.*
7. Reverse the spindle rotation until the cutting tool has just cleared the end of the threaded section.
8. Check the thread with a metric screw pitch gauge.
9. Cut the thread to the required depth.
   NOTE: *Never disengage the split nut until the thread has been cut to depth.*

To Reset a Threading Tool

A threading tool must be reset whenever it is necessary to remove partly threaded work and finish it at a later time, if the threading tool has had to be removed for regrinding, or if the dog slips on the work.

1. Mount the work and set up the lathe for thread cutting and for the proper number of threads to be cut.
2. With the threading toolbit clear of the work, start the lathe and engage the split-nut lever on the correct line for the pitch in millimetres or the number of threads per inch.
3. Allow the carriage to travel until the toolbit is opposite any portion of the unfinished thread.
4. Stop the lathe, but be sure to *leave the split-nut lever engaged.*
5. Feed the toolbit into the thread groove, using ONLY the *compound rest and crossfeed handles*, until the right-hand side of the toolbit touches the right-hand side of the thread (left side for left-hand threads) (Fig. 11-127).
6. Set the crossfeed graduated collar to zero *without moving the crossfeed handle*, and then set the compound rest feed collar to zero.
7. Back out the threading tool using the crossfeed handle, disengage the split-nut lever, and move the carriage until the toolbit clears the start of the thread.
8. Set the crossfeed handle back to zero and take a trial cut without setting the compound rest.
9. Set the depth of cut using the compound rest handle and take successive cuts to finish the thread.

Fig. 11-127
The right side of the threading tool should bear against the right side of the previously cut thread.

CHAPTER 11 / THE ENGINE LATHE

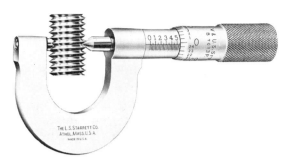

Fig. 11-128
A thread micrometer measures the pitch diameter of a thread.

THREAD MEASUREMENT

There are several methods of checking threads for depth, angle, and accuracy. Most commonly used are thread gauges, thread micrometers, and a finished master hexagon nut.

A finished master hexagon nut can be used for checking all general-purpose threads. The thread should be cut deep enough to allow the nut to turn on freely with no end play. Metric and inch thread micrometers can be used to check the pitch diameter of threads to an accuracy of 0.02 mm, or to 0.001 in. These micrometers are made to measure certain ranges of threads. Metric screw thread micrometers are available for checking thread diameters from 0 to 100 mm. Figure 11-128 shows a thread micrometer used to check 14 to 20 threads/in. (60°).

A thread ring gauge (Fig. 11-129) is often used in production work for testing threads. It is a hardened standard ring gauge that can be adjusted to compensate for wear or tolerance.

Fig. 11-129
A thread ring gauge used for checking threads on production work.

TEST YOUR KNOWLEDGE

1. Explain the cutting action of a lathe.
2. Name six operations which may be performed on a lathe.
3. How is the size of a lathe determined?

Lathe Parts
4. Name four parts of the headstock.
5. What is the purpose of the quick-change gearbox?
6. Name the three main parts of the carriage, and state one purpose for each.
7. For what purpose are the following used?
   (a) compound rest
   (b) apron handwheel
   (c) feed directional plunger
8. What purpose does the tailstock serve?

Safety Precautions
9. Why is loose clothing dangerous around machines?
10. Why should metal chips not be handled by hand?
11. Explain why heavy cuts should not be taken on long, slender pieces.

Cutting Speeds and Feeds
12. How is cutting speed defined?
13. What will result if the lathe speed is
    (a) too slow?
    (b) too fast?
14. Calculate the *r/min* required to take a rough cut from the following: (See Table 11-1 for the cutting speeds of various materials.)
    (a) A 75 mm piece of machine steel
    (b) A 44 mm piece of tool steel

    (c) A 3/4 in. diameter piece of machine steel
    (d) A 2-1/2 in. diameter piece of cast iron
15. Define lathe feed.
16. What feed, in both metric and inch units, is recommended for general purpose machining for:
    (a) rough cuts?
    (b) finish cuts?

Cutting Tools and Toolholders

17. Explain the difference between left-hand and right-hand toolbits.
18. What is the purpose of end relief and side relief on a toolbit?
19. Name three lathe toolholders, and state the purpose of each.
20. What is the advantage of using a turret type toolpost?
21. What is one advantage of the quick-change toolpost?

Work-Holding Devices

22. What purpose do lathe centres serve?
23. For what operations is a revolving tailstock centre particularly useful?
24. Name two types of chucks, and explain how each operates.
25. For what purpose are the following used?
    (a) collet chucks
    (b) mandrels
    (c) steady rests
    (d) follower rests
26. What is the advantage of a Jacobs chuck over a draw-in collet chuck?

Alignment of Lathe Centres

27. Why is it important that the headstock and tailstock centre be in line?
28. Name three methods of aligning lathe centres.
29. Briefly describe how to align the centres using the trial cut method.

Mounting and Removing Centres

30. Explain how the centre is removed from the headstock and the tailstock.
31. What should be done before replacing a centre in the headstock or tailstock?

Mounting Work Between Centres

32. How far should the toolholder extend beyond the toolpost?
33. How tight should the toolholder setscrew be tightened?
34. List six important steps involved in mounting work between centres.
35. How tight should the work be adjusted between centres?
36. List the sequence in which a piece of round work should be machined in a lathe.

Facing Work Between Centres

37. State three purposes of facing work.
38. Explain how the cutting tool must be set up for facing.
39. How should the cutting tool be fed in order to face a surface?

Graduated Micrometer Collars

40. What is the purpose of graduated micrometer collars?
41. What is the value of each graduation on the micrometer collar of lathes using:
    (a) SI?
    (b) the inch system of measurement?
42. If a 0.25 mm depth of cut is set on the crossfeed graduated collar, how much material would be removed from the diameter of a workpiece?
43. What rule should be followed when setting a depth of cut on machines where the work revolves?

Parallel Turning

44. When can a parallel diameter be produced on a lathe?
45. What is the purpose of a light trial cut before taking a rough or finish cut from a diameter?
46. How close to finished size should the rough cut be taken?
47. What feed should be used for rough turning?
48. What is the purpose of finish turning?
49. What feed should be used for finish turning?

50. How can you tell when the outside caliper setting is correct?

Shoulder Turning
51. Name three types of shoulders used in machine shop work.
52. To what length should the small diameter be turned when cutting a filleted shoulder?
53. How can short angular shoulders be cut?

Filing in a Lathe
54. Why is too much filing on a diameter not recommended?
55. How can damage to the lathe be prevented during filing and polishing?
56. List three safety precautions which should be observed when filing.

Polishing
57. What purpose does polishing serve?
58. Explain how the abrasive cloth should be held for polishing in a lathe.

Knurling
59. Define the process of knurling.
60. What is the purpose of knurling, and how is it performed?
61. Explain how the knurling tool should be set up for knurling.
62. If the knurling pattern is not correct, how can it be corrected?
63. Why should the lathe feed not be disengaged during the knurling operation?

Grooving
64. For what purpose are grooves used?
65. How can the depth of cut be gauged when grooving?

Tapers
66. Define a taper.
67. How are tapers generally expressed?
68. Name four common inch tapers and give the taper per foot for each.
69. A tapered pin has a taper of 3 mm in 120 mm. How would this taper be indicated on a drawing?
70. For three common inch tapers, state the amount of taper and where each is used.

Taper Calculations
71. Name four parts of a taper.
72. Calculate the taper per foot for the following:
    (a) Large diameter = 1.625 in.
        Small diameter = 1.425 in.
        Length of taper = 3 in.
    (b) Large diameter = 7/8 in.
        Small diameter = 7/16 in.
        Length of taper = 6 in.
73. (a) A metric taper pin has a taper of 1:50. If the small diameter of the pin is 6 mm and the pin is 75 mm long, what is the large diameter of the pin?
    (b) If the taper is 1:30, the small diameter is 9 mm, and the length is 105 mm, what is the large diameter of the pin?
74. (a) Calculate the tailstock offset for the problems in question 73 if:
    (i) the length of the work in problem (a) is 120 mm.
    (ii) the length of the work in problem (b) is 150 mm.
    (b) Calculate the tailstock offset for the problems in question 72 if:
    (i) the length of the work in problem (a) is 10 in.
    (ii) the length of the work in problem (b) is 9 in.
75. Calculate the tailstock offset for the following using the simplified tailstock offset formula for inch tapers.
    (a) L.D. = 3/4 in., S.D. = 17/32 in.,
        Length of taper = 6 in.,
        Overall length of work = 18 in.
    (b) L.D. = 7/8 in., S.D. = 25/32 in.,
        Length of taper = 3-1/2 in.,
        Overall length of work = 10-1/2 in.
76. Calculate the tailstock offset required to turn a 1:10 taper × 50 mm long on a workpiece 125 mm long. The small diameter of the tapered section is 50 mm.

Taper Turning
77. List four advantages of cutting a taper using a taper attachment.

78. What is the difference between a plain and a telescopic taper attachment?
79. How should the compound rest be set when cutting a taper on a plain taper attachment?
80. What adjustments must be made to offset a tailstock?
81. What type of tapers are generally cut with the compound rest?
82. How is the cutting tool fed when cutting tapers using the compound rest?

Checking a Taper
83. State two methods of checking a taper.
84. Describe how a taper may be checked by one of these methods.

Mounting and Removing Chucks
85. Why should a chuck cradle be used when mounting or removing a chuck?
86. Briefly explain how to remove a chuck from:
    (a) a taper spindle nose
    (b) a cam-lock spindle nose
87. When mounting a chuck on a threaded spindle nose, why should the chuck not be jammed against the shoulder?
88. Briefly explain how to mount a chuck on:
    (a) a taper spindle nose
    (b) a cam-lock spindle nose

Three-Jaw Chuck Work
89. For what purpose are three-jaw chucks used?
90. How far should the work extend beyond the jaws when facing in a chuck?
91. How can the compound rest be set to accurately face work to length?

Drilling and Reaming Holes
92. At what speed should the lathe be set for drilling centre holes?
93. Describe how the drill is mounted in a lathe to drill a 19 mm (or 3/4 in.) hole in a workpiece held in a chuck.
94. How can the depth of a hole be checked while it is being drilled on a lathe?
95. How much material should be left in holes up to 12 mm (1/2 in.) in diameter for reaming?
96. At what speed should the lathe be set for reaming?

Cutting Off and Tapping
97. How should the cutting-off tool be set in a lathe?
98. Name two things which will help avoid chatter during cutting off.
99. What type of tap should be used for tapping in a lathe?
100. How can the tap be guided to ensure that the thread will be true with the hole?

Standard Thread Forms
101. Define a thread.
102. List four thread forms, give the angle of each thread, and state a use for each.

Thread Terms and Calculations
103. Define the following thread terms:
    (a) major diameter
    (b) lead
    (c) pitch
    (d) root
104. (a) Calculate the depth, minor diameter, and width of flat for a 7/8 in. — 9 N.C. thread.
    (b) Calculate the pitch, depth, minor diameter, width of crest, and width of root for a M 16 × 3 thread.

To Set a Quick-Change Gearbox
105. What is the purpose of the quick-change gear box?

Thread-Chasing Dial
106. Explain the function of the thread-chasing dial.
107. What lines on the thread-chasing dial could be used for cutting 6, 11-1/2, and 16 threads/in.?

Thread Cutting
108. Explain how the threading toolbit is set up.

109. At what angle should the compound rest be set for cutting a 60° thread?
110. What should be the size of the diameter to be threaded?
111. At what position should the cross-feed handle be for threading?
112. What is the purpose of taking a light trial cut before threading the work?
113. Explain how the pitch in millimetres or the number of threads per inch can be checked.
114. How is the depth of each threading cut set?

To Convert an Inch-Designed Lathe to Metric Threading

115. Describe how a standard quick-change gear lathe may be set up to cut a metric thread.
116. What precaution must be taken when cutting a metric thread on a standard quick-change gear lathe?
117. What is the pitch, depth, minor diameter, width of crest, and width of root for a M 20 × 2.5 thread?

To Reset a Threading Tool

118. State two reasons why it may be necessary to reset a threading tool.
119. Explain in point form how to reset a threading tool in a partially cut thread.

Thread Measurement

120. List three methods used to check a thread.
121. What instrument can be used to measure the pitch diameter of a thread?

CHAPTER 12
THE SHAPER

Courtesy Cincinnati Shaper Co.

The shaper, developed in the mid-nineteenth century, was used for many years for producing flat surfaces. A single-point cutting tool, travelling in a horizontal plane with a reciprocating motion (back and forth), removed metal on each forward stroke only. Since the shaper used an interrupted cutting action, cemented carbide cutting tools were never used efficiently on a shaper. Therefore, it was not an efficient method of removing metal because it cut during only one-half the time that the cutting tool was moving.

With the development of the horizontal and vertical milling machines, the use of the shaper has continually declined. The milling machines, using multi-toothed cutters, have a continuous cutting action and remove metal five to ten times faster than the shaper. Cemented carbide cutting tools also are used successfully, and have increased production as much as one hundred times that which was possible with a shaper.

It is little wonder that the use of the shaper declined so drastically, and in the 1980s there are very few manufacturers, if any, in the world who still produce shapers. However, the basic operations of the shaper are included in this book because some school shops may still have a shaper, which provides a work station, and it is still possible to come across a shaper in some industries.

TYPES OF SHAPERS

The shaper is used to produce flat surfaces horizontally, on an angle, or in a vertical plane. The work may be held in a vise or fastened to the table. A single-point cutting tool is driven back and forth by a ram which travels in a horizontal plane with reciprocating action. The cutting tool peels off a chip from the work on each forward stroke only (Fig. 12-1).

The size of a shaper is determined by the largest cube which can be machined on it. For example, a 14 in. shaper can machine a block 14 in. × 14 in. × 14 in.; and a 300 mm shaper can machine a block 300 mm × 300 mm × 300 mm.

There are three types of shapers:
(a) the crank shaper,
(b) the gear shaper,
(c) the hydraulic shaper.

CRANK SHAPER

The crank shaper (Fig. 12-2) is the type most commonly used. Its ram is given its reciprocating motion by means of a rocker arm, which is operated by a crank pin from the main driving gear or "bull wheel". The ram carries the downfeed mechanism,

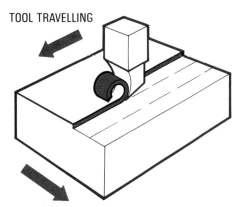

Fig. 12-1
The cutting action of a shaper.

which contains the clapper box and the cutting tool.

GEAR SHAPER

The gear shaper obtains the drive for the ram from a gear and a rack that is connected to the ram. This shaper also has a mechanism that provides a quick return for the ram.

HYDRAULIC SHAPER

The hydraulic shaper is driven by the movement of a piston in a cylinder of oil and is controlled by a valve mechanism connected with the oil pump. Its mechanical features are the same as those of the crank shaper.

SHAPER PARTS

To operate a shaper successfully, a knowledge of the main operative parts and their function is necessary.

The *ram* is a semi-cylindrical form of heavy construction that provides the forward and return strokes to the cutting tool. It contains the *ram positioning mechanism* and the tool head. The *ram adjusting shaft* is used to change the position of the stroke, and the *ram positioning lock* or *clamp* locks the ram in a fixed position.

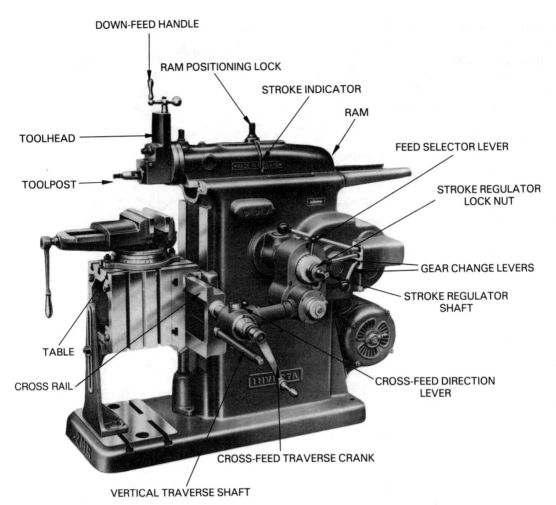

Fig. 12-2
The main parts of a crank shaper.

Courtesy Elliot Machine Tools

The *toolhead* (Fig. 12-3) is fastened to the ram to hold the toolholder and the *clapper box*, which allows the cutting tool to rise slightly on the return stroke. The *downfeed handle* provides a means of feeding the cutting tool to any given dimension or depth of cut in either hundredths of a millimetre or thousandths of an inch, as measured on its graduated collar.

The *crossfeed traverse crank* is used to provide a horizontal movement to the table in either hundredths of a millimetre or thousandths of an inch with the use of the graduated collar. This screw is used to move the table longitudinally under the cutting tool. The *crossfeed direction lever* may be engaged in a ratchet to provide an automatic feed to the table. The *vertical traverse shaft* is used to lower or raise the table.

The *table* is fastened to the *crossrail*, and it provides a support for the work to be machined. It can be raised or lowered by the elevating screw, and moved away from or toward the operator by hand with the crossfeed screw, or by power with the automatic feed mechanism.

188 CHAPTER 12 / THE SHAPER

The *stroke regulator shaft* adjusts the length of stroke required, and the *stroke regulator lock nut* is used to lock the mechanism in a fixed position. The *stroke indicator* provides a guide when adjusting the shaper to the proper length of stroke required.

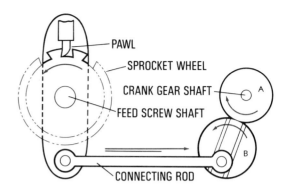

Fig. 12-4
The table feed mechanism of a crank shaper.

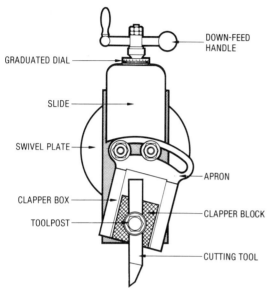

Fig. 12-3
The parts of a shaper toolhead.

SHAPER FEED MECHANISM

When horizontal surfaces are being machined, the table upon which the work is mounted feeds the work toward the cutting tool automatically *on the return stroke of the ram*. When vertical feeds are required, the feed is controlled by turning the downfeed screw handle by hand or automatically if the shaper is equipped with an automatic downfeed mechanism, on each return stroke of the ram.

Figure 12-4 shows the basic principle on which the table feed mechanism operates on a crank-type shaper. Gear A is fastened to the crank gear shaft. When the crank gear turns to produce the return stroke of the ram, gear A turns 140°. This turns gear B 140° and moves the connecting rod in the direction of the arrow. This causes the pawl to turn the sprocket wheel, moving the table on the return stroke of the ram. The further that the connecting rod is set off centre in Gear B causes the pawl to travel a greater distance and produces a coarser crossfeed.

To Engage or Disengage the Table Feed
1. Move the cutting tool clear of the work.
2. Turn the pawl knob until the pawl engages in the sprocket wheel as in Fig. 12-4.
3. Start the shaper.
4. Check the position of the connecting rod in the T-slot in gear B to see that the feed operates on the return stroke of the ram.
5. To set the amount of feed, adjust the connecting rod off centre at B so that the pawl will move (index) two or more teeth on the sprocket wheel *when the ram is on its return stroke*.
6. To disengage the feed raise the pawl knob and turn the pawl 90° to prevent the pawl from entering the teeth in the sprocket wheel.

SHAPER SPEEDS

Speeds and feeds are dependent upon many factors that control the efficiency of

TABLE 12-1 SHAPER SPEEDS AND FEEDS

Cutting Tool	Machine Steel				Tool Steel				Cast Iron				Brass			
	Speed per min		Feed		Speed per min		Feed		Speed per min		Feed		Speed per min		Feed	
	m	ft.	mm	in.	m	ft.	mm	in.	m	ft.	mm	in.	m	ft.	mm	in.
H.S.S.	24	80	0.25	0.010	15	50	0.38	0.015	18	60	0.51	0.020	48	160	0.25	0.010
Carbide	46	150	0.25	0.010	46	150	0.30	0.012	30	100	0.30	0.012	92	300	0.38	0.015

the shaper use. Factors such as depth of cut, the amount of feed, the material to be machined, and the type of cutting tool control the speeds and feeds for machining. Table 12-1 gives recommended shaper speeds and feeds.

FORMULA FOR CALCULATING STROKES PER MINUTE

Inch Speed Calculations

FORMULAS: To obtain number of strokes per minute.

$$N = \frac{CS \times 7}{L}$$

N = number of strokes per minute
L = length of work plus 1 in. for tool clearance
CS = cutting speed of metal in feet per minute

EXAMPLE: How many strokes per minute should the ram be set to shape a piece of machine steel 9 in. long?

$$N = \frac{CS \times 7}{L}$$
$$= \frac{80 \times 7}{10}$$
$$= 56 \text{ strokes/minute}$$

Metric Speed Calculations

When shaping a workpiece with metric dimensions, set the length of stroke to 3 cm longer than the workpiece, to allow for clearance at each end of the stroke. The number of strokes per minute is calculated as follows:

Strokes per minute (N) =
$$\frac{\text{Cutting speed in metres}}{\text{Length of strokes in metres}} \times \frac{3}{5}$$

$$N = \frac{CS \text{ (metres)}}{\text{Length of stroke (metres)}} \times 0.6$$

EXAMPLE: How many strokes per minute will be required to machine a piece of tool steel 33 cm long (CS 15)?

Length of stroke (L) = 33 + 3 cm

$$N = \frac{CS}{L} \times 0.6$$
$$= \frac{15}{0.36} \times 0.6$$
$$= 25 \text{ strokes/minute}$$

CUTTING TOOLS

Shaper cutting tools are similar to lathe cutting toolbits. The shaper is frequently called upon to make many different types of cuts on flat surfaces. For this reason, there are shaper cutting tools ground to provide a cutting edge for cutting horizontal right or left, for cutting vertical right or left, and for grooving. All these tools must have side-relief, end-relief, and side-rake angles in order to cut properly.

The usual practice in shaping a flat surface is to take rough and finish cuts. The rough cut can be taken with coarse feeds, leaving the surface about 0.80 mm (or 1/32 in.) oversize to allow for a finishing cut. Some of the factors that frequently cause chatter marks are: too high a cutting speed, poor setting of the tool, insecure clamping of the work, incorrect cutting angle of the tool, too much cutting tool overhang.

Figure 12-5 shows the setting of the cutting tool for machining. The cutting tool is well supported in the toolpost to ensure the minimum spring to the cutting edge. Also, the work is as close as possible to the toolhead so that rigidity is maintained.

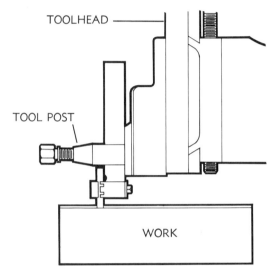

Fig. 12-5
A short overhang of the tool slide will provide rigidity and prevent chattering.

WORK-HOLDING DEVICES

Since the size and shape of workpieces machined in a shaper vary greatly, a variety of devices are used to hold the workpiece. Some of the more common devices used to support or hold the work are the vise, parallels, hold-downs, and clamps and bolts.

SHAPER VISE
Most of the work machined on a shaper is held in a vise that is fastened to the table. The shaper vise (Fig. 12-6) consists of a body or base that is graduated in degrees

Courtesy Kostel Enterprises Ltd.

Fig. 12-6
The standard vise on a shaper.

and can be swivelled to any angle. Most work machined in a vise is either parallel or at right angles to the ram. This vise has a movable and a fixed jaw. The movable jaw is heavier than the fixed jaw because the cut is seldom taken against the fixed jaw.

PARALLELS
Parallels (Fig. 12-7) are square or rectangular bars of steel or cast iron with opposite sides parallel and adjacent sides square. They are made in pairs of various sizes and lengths, depending on their use. For precision work, parallels should be hardened and ground. They are used to raise the work to the required height in the vise level to allow the machining to be done above the vise jaws.

Courtesy Taft-Peirce Manufacturing Co.

Fig. 12-7
Parallels are used to raise the work in a vise and provide a solid seat.

Fig. 12-8
The graduated micrometer collar on the toolhead allows an accurate depth of cut to be set.

GRADUATED MICROMETER COLLAR

The graduated collar on the downfeed screw allows the operator to accurately set the depth of cut to be taken in either hundredths of a millimetre or in thousandths of an inch (Fig. 12-8). Since the work on a shaper is clamped and the cutting tool reciprocates, all cuts are taken from one surface of the work. Therefore, when setting the depth of cut, always set the graduated collar for the amount of material to be removed.

Inch collars are graduated in 0.001 in. increments. To set the depth of cut to remove 1/32 in. from a surface, calculate or refer to a decimal equivalent chart and find the reading to the nearest thousandth which is 0.031 in. Set the graduated collar at zero and turn the downfeed handle until the graduated collar shows 31 graduations. The cutting tool will then remove 1/32 in. from the surface.

Metric collars are usually graduated in 0.02 mm increments. In order to remove 0.80 mm from the work, the downfeed handwheel is turned until 40 graduations show on the collar.

SAFETY PRECAUTIONS

It is very important that the operator be familiar with the various areas of a shaper which could be hazardous. By following a few basic safety precautions and using common sense, most accidents can be avoided.

1. Always wear safety glasses to protect your eyes from flying chips (Fig. 12-9).
2. Never operate any machine until you have been properly instructed.
3. *Before starting* a shaper, make sure that the work, vise, tool, and ram are securely fastened.
4. Check that the tool and toolhead will clear the work and also clear the column on the return stroke.
5. Always prevent chips from flying by using a metal shield or wire screen. This could prevent injury to others.

Courtesy Kostel Enterprises Ltd.

Fig. 12-9
Always wear approved safety glasses when operating a shaper.

MOUNTING THE WORKPIECE

In order to produce accurate work on a shaper, it is important that the workpiece be mounted accurately and securely. If the work is not mounted securely, the workpiece may be thrown out of the machine and cause a serious accident.

To Mount Work in a Shaper Vise
1. Remove all burrs from the workpiece.
2. Clean the work and the vise thoroughly.
3. Select a set of parallels large enough to raise the work about 6 mm (1/4 in.) above the vise jaws.
4. Set the work on the parallels in the centre of the vise.
5. Place a short paper feeler between each corner of the work and the parallels (Fig. 12-12).
6. *Tighten the vise securely.* If the vise is not tightened securely, when the work is tapped down, it will hit the parallels and bounce up.
7. Tap the work down onto the parallels with a soft-faced hammer, alternating the hammer blows between each end of the work.
8. Tap each corner of the work down until all the paper feelers are tight.
9. *DO NOT retighten the vise* after the work has been tapped down, because the work will be forced up and off the parallels during the tightening process.

Courtesy Kostel Enterprises Ltd.

Fig. 12-10
Always stand to one side of the cutting stroke to prevent injury from flying chips.

6. *Always* stand parallel to the cutting stroke and *not* in front of it (Fig. 12-10).
7. Never attempt to remove chips or reach across the table while the ram is in motion.
8. Keep the area around a machine neat and tidy (Fig. 12-11).
9. Clean oil and grease from the floor immediately to prevent dangerous falls.
10. Never attempt to adjust a machine while it is in motion.

Courtesy Kostel Enterprises Ltd.

Fig. 12-11
A dirty, cluttered floor may result in a dangerous fall.

Fig. 12-12
The proper method of mounting work in a shaper vise.

MOUNTING THE WORKPIECE 193

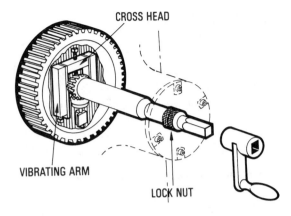

Courtesy Kostel Enterprises Ltd.

Fig. 12-13
The shaper stroke regulating mechanism.

SETTING THE SHAPER STROKE

The shaper stroke must be set to suit the length of each workpiece, and also the position where it is held in a vise, on the table, or other work-holding device. If the shaper stroke is longer than necessary, valuable time will be wasted since it will take longer to complete the job.

To Set the Length of the Shaper Stroke

1. Mount the work in the shaper vise.
2. Measure the length of the work and add 25 mm (or 1 in.) for tool clearance (12 mm or 1/2 in. at each end of the stroke).
3. Use the start-stop or inching button to jog the motor until the ram is at the extreme back end of the stroke.
   OR
   Place the crank on the stroke regulator shaft and turn it until the ram is at the back end of the stroke.
4. Loosen the *stroke regulator lock-nut* about one-half turn (Fig. 12-13).
5. Turn the stroke regulator shaft until the *stroke indicator* is at the desired length of stroke.
6. Tighten the stroke regulator lock-nut. In Figure 12-13, the crosshead that fits into the vibrating arm is the part moved by the stroke regulator screw, away from the centre to lengthen the stroke, and toward the centre to shorten it.

To Set the Position of the Shaper Stroke

Before setting the position of the shaper stroke, the length of stroke must be first set as in the previous operation. The ram must then be adjusted so that the cutting tool will machine the full length of the piece.

1. Set the length of the stroke as in the previous operation.
2. Be sure that the ram is at the extreme back end of the stroke.
3. Loosen the ram lock or clamp (Fig. 12-14).
4. Pull the toolhead and ram forward until the toolbit is within 12 mm (1/2 in). of the work (Fig. 12-14).
   OR
   If the ram is equipped with an adjusting screw, turn it until the toolbit is within 12 mm (1/2 in.) of the work.

Courtesy Kostel Enterprises Ltd.

Fig. 12-14
Setting the position of the shaper stroke.

5. Tighten the ram lock or clamp.
6. With the cutting tool clear of the work, start the machine and see that the toolbit clears each end of the work by about 12 mm (1/2 in.)

Courtesy Kostel Enterprises Ltd.

Fig. 12-15
Swinging the apron so that the cutting tool does not rub on the return stroke.

SHAPING A FLAT SURFACE

Most work machined in a shaper is held in a vise. After the work is properly set up in the vise the cutting tool must be set to the surface of the work, and then the surface is machined.

1. Mount the work in the vise as in the operation described in the section "To Mount Work in a Shaper Vise" in this chapter.
2. Set the length and postion of the shaper stroke.
3. Loosen the apron lock screw and swing the top of the toolhead apron to the left (Fig. 12-15).
   This will prevent the toolbit from rubbing on the return stroke.
4. Tighten the apron lock screw and set the toolholder vertically (Fig. 12-15).

To Set the Toolbit to the Work

5. Start the shaper, and stop it when the toolbit is over the work.
6. Hold a piece of paper between the toolbit and the surface of the work.
7. Turn the downfeed handle until a slight drag is felt on the paper (Fig. 12-16).
8. Loosen the graduated collar lock screw and set the collar to zero.
9. Tighten the graduated collar lock screw.
10. Move the table until the cutting tool clears the edge of the work by about 6 mm (1/4 in.).

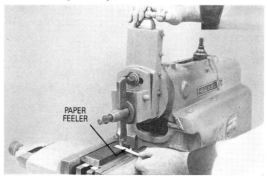

Courtesy Kostel Enterprises Ltd.

Fig. 12-16
Setting the cutting tool to the surface of the work.

To Machine the Surface

11. Set the depth of cut by turning the down-feed handle *clockwise* the required amount (about 1.5 mm or 0.060 in. to clean up a surface).
12. Set the shaper speed and feed for the material being cut. (See Table 12-1).
13. Start the shaper and engage the automatic feed, being sure that the feed operates on the return stroke of the ram.
14. Machine the surface and then remove all burrs from the edges of the workpiece with a file.

MACHINING A BLOCK SQUARE AND PARALLEL

In order to machine the four sides of a piece of work so that its sides are square and parallel, it is important that each side be machined in a definite order (Fig. 12-17). It is very important that dirt and burrs be removed from the work, vise, and parallels, since they can cause inaccurate work.

See the operation of Machining a Block Square and Parallel in the Vertical Mill section of Chapter 13 for the detailed sequence to follow when setting up work for machining the four sides of a block. The setup sequence for both machines is exactly the same.

Courtesy Kostel Enterprises Ltd.

Fig. 12-17
The sequence for machining the sides of a rectangular piece square and parallel.

MACHINING A VERTICAL SURFACE

A vertical surface can be machined on flat work by feeding the toolhead with the down-feed handle. In order for the machined edge to be square, the toolhead must be set at 90° to the top of the shaper table.

1. Align the toolhead with a square.
2. Set the shaper vise at 90° to the ram.

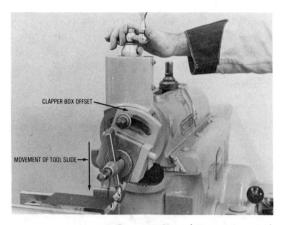

Courtesy Kostel Enterprises Ltd.

Fig. 12-18
The clapper box must be offset when machining a vertical surface.

3. Place a suitable pair of parallels in the vise.
4. Adjust the work so that about 12 mm (1/2 in.) extends past the end of the vise jaws and parallels (Fig. 12-18).
5. Tighten the vise securely and then tap the work down until the paper feelers are tight.
6. Adjust the length and position of the shaper stroke.
7. Swing the top of the apron *away from the surface to be cut* (Fig. 12-18).
8. Fasten a left-hand toolholder in the toolpost.
9. Mount a facing tool in the toolholder.
10. Swivel the toolholder so that it will clear the side of the surface to be cut.
11. Raise the toolhead so that the toolbit clears the top of the work by about 6 mm (1/4 in.).
12. Set the depth of cut by turning the crossfeed traverse crank.
13. Set the shaper for the proper speed and then start the shaper.
14. Feed the toolbit down by turning the downfeed handle clockwise about 0.25 mm (or 0.010 in.) on each return stroke of the ram.
15. If necessary, take further cuts to bring the work to size.

MACHINING AN ANGULAR SURFACE

Angular surfaces may be cut on a shaper by three methods:
(a) setting the workpiece on the desired angle,
(b) swivelling the vise to the required angle,
(c) setting the toolhead to an angle.

A

B

Courtesy Kostel Enterprises Ltd.

Fig. 12-19
Using a rule to check the distance from the top of the vise to the layout line.

To Set Up Work for Angular Shaping
1. Set the work in a vise with the layout line approximately 6 mm (1/4 in.) above the vise jaw.
2. Align the layout line parallel to the top of the vise jaw by:
   (a) laying a parallel on top of the vise and tapping the work until the layout line is even with the top of the parallel.
   OR
   (b) checking the line at each end of the work with a surface gauge.
   OR
   (c) measuring the distance from the top of the vise jaw to the layout line at each end of the work with a rule. Tap the work until the measurement at both ends is the same (Fig. 12-19).
3. Adjust the length and position of the shaper stroke.
4. Machine the angular surface by horizontal shaping (Fig. 12-20).

Courtesy Kostel Enterprises Ltd.

Fig. 12-20
Machining an angular surface by horizontal shaping.

TEST YOUR KNOWLEDGE

1. How are shapers designated for size?
2. Name three types of shapers, and state how the ram of each is driven.

Shaper Parts

3. List the parts that are fastened to the ram.
4. What part of the toolhead is used to set the depth of cut?
5. Name the part that
   (a) operates the table horizontally
   (b) adjusts the ram
   (c) adjusts the length of stroke
   (d) raises and lowers the table
6. Explain how the table feed mechanism operates.

Shaper Speeds

7. What factors affect the shaper speeds and feeds?
8. How is shaper speed expressed in both metric and inch systems?
9. What speeds and feeds are recommended in metric and inch units for:
   (a) cast iron?
   (b) machine steel?
   (c) brass?
   (d) tool steel?
10. Calculate the number of strokes per minute to machine a piece of:
    (a) cast iron 7 in. long
    (b) machine steel 12 in. long
    (c) tool steel 55 mm long
    (d) brass 410 mm long

Cutting Tools

11. List four factors which can cause chatter during machining.
12. Why is too much overhang of the toolhead not advisable?

Work-Holding Devices

13. Name three methods of holding work for a shaping operation.
14. For what purpose are parallels used?

Graduated Micrometer Collar

15. What is the main purpose of a graduated collar?
16. Explain why the depth of cut set on a shaper is always the amount to be removed.
17. What is the value of each line
    (a) on a metric graduated collar?
    (b) on an inch graduated collar?

Safety Precautions

18. List four important safety precautions, and explain why each should be observed.

Mounting the Workpiece

19. How high should work be set above the vise jaws?
20. What is the purpose of using paper feelers between the parallels and the work?
21. Why should the vise not be retightened after the work has been tapped down?

Setting the Shaper Stroke

22. How long should the shaper stroke be in relation to the length of the workpiece? Explain why.
23. In what position must the ram be before adjusting the shaper stroke?
24. Explain how the shaper stroke is set to length.
25. Explain how the position of the shaper stroke is set.

Shaping a Flat Surface

26. Why should the toolhead apron be swung to the left?
27. How is the toolbit set to the surface of the work?
28. How much must be generally removed in order to clean up a surface?
29. On what stroke should the feed operate?

Machining a Vertical Surface

30. In what position should the toolhead be for machining a vertical surface?
31. Explain how the work should be set in a vise for machining a vertical surface.
32. How much should the toolbit be fed on each return stroke of the ram?

Machining an Angular Surface

33. Name three methods that can be used to cut an angular surface on a shaper.
34. Explain how the workpiece can be set in a vise in order to cut an angular surface.

CHAPTER 13
MILLING MACHINES

Courtesy Cincinnati Milacron Inc.

The milling machine is a machine tool used to produce accurately machined surfaces such as flats, angular surfaces, grooves, cams, contours, gear and sprocket teeth, helical grooves, and accurately sized holes. These operations are performed by feeding the workpiece into a revolving multi-toothed cutter. The shape of the milling cutter will determine the shape of the finished surface.

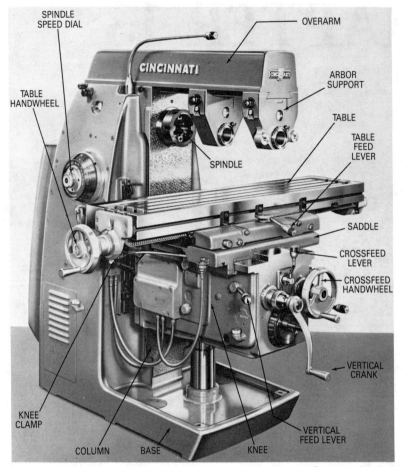

Fig. 13-1
Courtesy Cincinnati Milacron Inc.

The universal knee and column type milling machine.

The versatility of the milling machine makes it suitable for production, tool room, jobbing shops, and experimental and research work. The more commonly used milling machines are:
(a) plain knee and column
(b) universal knee and column (Fig. 13-1)
(c) vertical knee and column
(d) the manufacturing types
(e) automation type.

The *universal* knee and column type is perhaps the most versatile and can be adapted to perform many jobs by the use of a variety of attachments. The *plain* knee and column milling machine is the same as the universal, except that it does not have the swivel table housing.

PARTS OF THE MILLING MACHINE

The *base* gives support and rigidity to the machine and also acts as a reservoir for the cutting fluids.

The *column face* is a precision-machined and scraped section used to support and guide the knee when it is moved vertically.

The *knee* is attached to the column face and may be moved vertically on the column face either manually or automatically. It generally houses the feed mechanism.

The *saddle* is fitted on top of the knee and may be moved in or out manually by means of the crossfeed handwheel or automatically by the crossfeed engaging lever.

The *swivel table housing*, fastened to the saddle on a universal milling machine, enables the table to be swivelled 45° to either side of the centre line.

The *table* rests on guideways in the saddle and travels longitudinally in a horizontal plane. It supports the vise and the work.

The *crossfeed handwheel* is used to move the table toward or away from the column.

The *table handwheel* is used to move the table horizontally back and forth in front of the column.

The *feed dial* is used to regulate the table feeds.

The *spindle* provides the drive for arbors, cutters, and attachments used on a milling machine.

The *overarm* provides for correct alignment and support of the arbor and various attachments. It can be adjusted and locked in various positions, depending on the length of the arbor and the position of the cutter.

The *arbor support* is fitted to the overarm and can be clamped at any location on the overarm. Its purpose is to align and support various arbors and attachments.

The *elevating screw* is controlled by hand or an automatic feed. It gives an upward or downward movement to the knee and the table.

The *spindle speed dial* is set by a crank that is turned to regulate the spindle speed. On some milling machines the spindle speed changes are made by means of two levers. When making speed changes, always check to see whether the change can be made when the machine is running, or if it must be stopped.

Most milling machines are equipped with two or more arbors on which many different shaped cutters (Fig. 13-2) may be mounted. Most milling machines can be fitted with many other attachments, such as a dividing head, a rotary table, a vertical head, a slotting attachment, a rack-cutting attachment, and various special fixtures.

MILLING CUTTERS

A milling cutter is a rotary cutting tool having equally spaced teeth around the periphery and sometimes on the end or sides. These teeth engage with the workpiece to remove the metal in the form of chips. Milling cutters are manufactured in a variety of shapes and sizes to produce many shapes or profiles. They are usually made of high-speed or special steel with tungsten carbide teeth. Some types of common cutters are as follows:

Plain Milling Cutters are the most commonly used type of milling cutter. They are usually wide cylindrical cutters with the teeth on the periphery. Plain milling cutters are used for producing flat surfaces such as light duty, light duty helical, heavy duty, and high helix (Fig. 13-2A).

Side Milling Cutters are comparatively narrow cylindrical cutters having teeth on the sides and the periphery. The teeth may be straight or staggered (Fig. 13-2B), where each tooth is alternately set to the right and to the left for better chip clearance. Side milling cutters are used for facing the edges of work and cutting slots.

Angular Cutters have teeth which are at an angle to the face and the axis of the cutter. Single angle cutters and double angle cutters (Fig. 13-2C) are used for milling angular surfaces, grooves, and serrations.

Formed Cutters are the exact shape of the part to be produced which permits duplication of parts more economically than by most other means. Formed cutters

A *Courtesy Cincinnati Milacron Inc.*

C *Courtesy Cleveland Twist Drill (Canada) Ltd.*

E *Courtesy Butterfield Division, Union Twist Drill Co.*

B *Courtesy DoAll Company*

D *Courtesy Butterfield Division, Union Twist Drill Co.*

F *Courtesy Weldon Tool Co.*

Fig. 13-2
A few of the many types of cutters commonly used in milling practice.

may be of practically any shape such as concave, convex, and irregular. A convex cutter is shown in Fig. 13-2D.

Gear Cutters (Fig. 13-2E) are another type of formed cutter. They are manufactured in a wide range of sizes and contours, depending on the number of teeth and the pitch of the gear. They are generally used for special gear needs, since most gears are now mass produced by hobbing.

Shell End Mills (Fig. 13-2F) have teeth on the periphery and on the end. They can be used for facing and peripheral cutting. They are held on a stud arbor, which fits into and is driven by the milling machine spindle. The cutters may be threaded onto the arbor or held on by a special type of cap screw and driven by a key.

MILLING MACHINE SAFETY

The milling machine, like any other machine, demands the total attention of the operator and a thorough understanding of the hazards involved in its operation. The following rules should be observed when operating the milling machine.

Courtesy Kostel Enterprises Ltd.

Fig. 13-3
Always use a cloth when handling milling cutters.

1. Be sure that the work and the cutter are mounted securely before taking a cut.
2. Always wear safety glasses.
3. When mounting or removing milling cutters, always hold them with a cloth to avoid being cut (Fig. 13-3).
4. While setting up or measuring work, move the table as far as possible from the cutter to avoid cutting your hands (Fig. 13-4).
5. *Never* attempt to mount, measure, or adjust work until the cutter is *completely* stopped.
6. Keep hands, brushes, and rags away from a revolving milling cutter *at all times*.
7. When using milling cutters, do not use an excessively heavy cut or feed.
   NOTE: This can cause a cutter to break and the flying pieces could cause serious injury.
8. Always use a *brush*, not a rag, to remove the cutting after the cutter has stopped revolving (Fig. 13-5).
9. Never reach over or near a revolving cutter; keep hands at least 300 mm (12 in.) from a revolving cutter.
10. Keep the floor around the machine free of chips, oil, and cutting fluid (Fig. 13-6).

Courtesy Kostel Enterprises Ltd.

Fig. 13-5
Use a brush, not your hand, to remove chips.

Courtesy Kostel Enterprises Ltd.

Fig. 13-4
Stop the machine and move the work well clear of the cutter before measuring work.

Courtesy Kostel Enterprises Ltd.

Fig. 13-6
Keep the floor around the machine clean and free of hazards.

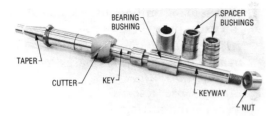

Courtesy Kostel Enterprises Ltd.

Fig. 13-7A
A milling machine arbor holds and drives the cutter.

MOUNTING AND REMOVING AN ARBOR

The milling arbor (Fig. 13-7A), which is used to hold the cutter during the milling operation, is held in the spindle by the draw-in bar (Fig. 13-7B). The cutter is driven by a key which fits into the keyways on the arbor and the cutter to prevent it from turning on the arbor. Spacer and bearing bushings hold the cutter in position on the arbor. When mounting or removing an arbor, follow the proper procedure to preserve the accuracy of the machine.

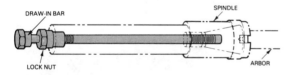

Fig. 13-7B
The draw-in bar holds the arbor firmly in the spindle.

To Mount an Arbor

1. Clean the tapered hole in the spindle and the taper on the arbor, using a clean cloth.
2. Check that there are no cuttings or burrs in the taper, which would prevent the arbor from running true.
3. Mount the tapered end of the arbor into the spindle taper.
4. Place the right hand on the draw-in bar (Fig. 13-8), and turn the thread into the arbor approximately 25 mm (1 in.)
5. Tighten the draw-in bar lock nut securely against the back of the spindle (Fig. 13-9).

Courtesy Kostel Enterprises Ltd.

Fig. 13-8
Inserting an arbor into the machine spindle.

Courtesy Kostel Enterprises Ltd.

Fig. 13-9
Tighten the draw-in bar locknut to hold the arbor securely in the spindle.

Courtesy Kostel Enterprises Ltd.

Fig. 13-10
Releasing the arbor taper by tapping the draw-in bar with a soft-faced hammer.

To Remove an Arbor
1. Remove the milling machine cutter.
2. Loosen the lock nut on the draw-in bar approximately two turns (Fig. 13-7B).
3. With a soft-faced hammer, hit the end of the draw-in bar until the arbor taper is free in the spindle (Fig. 13-10).
4. With one hand hold the arbor, and unscrew the draw-in bar from the arbor with the other hand (Fig. 13-11).
5. Carefully remove the arbor from the tapered spindle so as not to damage the spindle or arbor tapers.
6. Leave the draw-in bar in the spindle for further use.

Courtesy Kostel Enterprises Ltd.

Fig. 13-11
Holding the arbor and unscrewing the draw-in bar.

Courtesy Kostel Enterprises Ltd.

Fig. 13-12
Positioning the cutter on the arbor with the required number of spacer bushings.

MOUNTING AND REMOVING A MILLING CUTTER

Milling cutters must be changed frequently, so it is important that the following sequence be followed in order not to damage the cutter, the machine, or the arbor.

To Mount a Milling Cutter
1. Remove the arbor nut and collars.
2. Clean all surfaces of cuttings and burrs.
3. Check the direction of the arbor rotation.
4. Slide the collars on the arbor to the position desired for the cutter (Fig. 13-12).

Courtesy Kostel Enterprises Ltd.

Fig. 13-13
Sliding the arbor support into place over the bearing bushing.

5. Hold the cutter with a cloth and slide it on to the arbor fitted with a key.
6. Slide the arbor support in place and be sure that it is on a bearing bushing on the arbor (Fig. 13-13).
7. Put on additional spacers leaving room for the arbor nut.
   TIGHTEN THE NUT BY HAND.
8. Lock the overarm in position.
9. Tighten the arbor nut firmly with a wrench.
10. Lubricate the bearing collar in the arbor support.
11. Make sure that the arbor and arbor support will clear the work (Fig. 13-14).

Courtesy Kostel Enterprises Ltd.

Fig. 13-15
Be sure that the arbor support is supporting the bearing bushing before tightening the arbor nut with a wrench.

4. Loosen the arbor support and remove it from the overarm.
5. Remove the nut, spacers and cutter. Place them on a board (Fig. 13-16), not on the table surface.
6. Clean the spacer and nut surfaces and replace them on the arbor.
   Do not use a wrench to tighten the arbor nut.
7. Place the cutter in the proper storage.

Courtesy Kostel Enterprises Ltd.

Fig. 13-14
Before taking a cut, be sure that the arbor support will clear the work and/or the vise.

Removing a Milling Cutter
1. Be sure that the arbor support is in place and supporting the arbor on a bearing bushing before using a wrench on the arbor nut. This will prevent bending the arbor (Fig. 13-15).
2. Clean all cuttings from the arbor and cutter.
3. Loosen the arbor nut, using a wrench.

Courtesy Kostel Enterprises Ltd.

Fig. 13-16
A piece of masonite is used to protect the milling machine table when changing a cutter.

WORK-HOLDING DEVICES

There are several devices used in industry for holding work to be milled. The most commonly used are: vise, V-blocks, strap clamps, angle plates, and fixtures.

The *vise* can be used for holding square, round, and rectangular pieces for the cutting of keyways, grooves, flat surfaces, angles, gear racks, and T-slots.

V-blocks usually have a 90° V-shaped groove and a tongue which fits into the table slot to allow proper alignment for milling special shapes, flat surfaces, or keyways in round work.

Angle plates are used for holding large work or special shapes when machining one surface square with another.

A *fixture* (Fig. 13-17) is a special holding device made to hold a particular workpiece for one or more milling operations on a production basis. It provides an easy setup method but is limited to the job for which it is made.

Courtesy Cincinnati Milacron Inc.

Fig. 13-17
Milling a workpiece held in a fixture.

CUTTING SPEEDS

The cutting speed for a milling cutter is the speed, in either metres per minute or feet per minute, that the periphery of the cutter should travel when machining a certain metal. The speeds used for milling machine cutters are much the same as those used for any cutting tool. Several factors must be considered when setting the *r/min* to machine a surface. The most important are the:

- material to be machined
- type of cutter
- finish required
- depth of cut
- rigidity of the machine and the workpiece.

It is good practice to start from the calculated *r/min*, using Table 13-1, and then progress until maximun tool life and economy is accomplished. The speed of a milling machine cutter is calculated as for a twist drill or a lathe workpiece.

Inch Cutting Speeds

FORMULA: $r/min = \dfrac{CS \times 4}{D}$

	TABLE 13-1 MILLING MACHINE CUTTING SPEEDS			
	High-Speed Steel Cutter		Carbide Cutter	
Material	m/min	ft./min	m/min	ft./min
Machine steel	21-30	70-100	45-75	150-250
Tool steel	18-20	60-70	40-60	125-200
Cast iron	15-25	50-80	40-60	125-200
Bronze	20-35	65-120	60-120	200-400
Aluminum	150-300	500-1000	150-300	1000-2000

TABLE 13-2 RECOMMENDED FEED PER TOOTH
(High-Speed Steel Cutters)

Material	Side Mills		End Mills		Plain Helical Mills		Saws	
	millimetres	inches	millimetres	inches	millimetres	inches	millimetres	inches
Machine Steel	0.18	0.007	0.15	0.006	0.25	0.010	0.05	0.002
Tool Steel	0.13	0.005	0.10	0.004	0.18	0.007	0.05	0.002
Cast Iron	0.18	0.007	0.18	0.007	0.18	0.010	0.05	0.002
Bronze	0.20	0.008	0.23	0.009	0.28	0.011	0.08	0.003
Aluminum	0.33	0.013	0.28	0.011	0.46	0.018	0.13	0.005

EXAMPLE: Find the *r/min* to mill a keyway 1 in. wide in a machine steel shaft, using a 4 in. high-speed steel cutter. (CS 80)

$$r/min = \frac{CS \times 4}{D}$$
$$= \frac{80 \times 4}{4}$$
$$= 80$$

EXAMPLE: Calculate the *r/min* required to end-mill a 3/4 in. wide groove in aluminum with a high-speed steel cutter. (CS 600)

$$r/min = \frac{CS \times 4}{D}$$
$$= \frac{600 \times 4}{3/4}$$
$$= 3200$$

Metric Cutting Speeds

For metric cutters, the cutting speed is expressed in metres per minute. The formula used to determine the *r/min* is

$$r/min = \frac{CS \times 320}{D}$$

EXAMPLE: At what *r/min* should a 150 mm carbide tipped milling cutter revolve to machine a piece of machine steel? (CS 60)

$$r/min = \frac{CS \times 320}{D}$$
$$= \frac{60 \times 320}{150}$$
$$= 128$$

EXAMPLE: Calculate the *r/min* required to mill a groove 25 mm wide in a piece of aluminum using a high-speed end mill. (CS 150).

$$r/min = \frac{CS \times 320}{D}$$
$$= \frac{150 \times 320}{25}$$
$$= 1920$$

MILLING FEEDS

Feed is the rate at which the work moves into the revolving cutter, and it is measured either in millimetres per minute or in inches per minute. *Chip per tooth* is the amount of material which should be moved by each tooth of the cutter as it revolves and advances into the work. The *milling feed* is determined by multiplying the chip size (chip per tooth) desired, the number of teeth in the cutter, and the *r/min* of the cutter.

Inch Calculations

The formula used to find work feed in inches per minute is:

$$\text{feed} = N \times \text{c.p.t.} \times r/min$$

where:
- N = the number of teeth in the milling cutter
- c.p.t. = chip per tooth for a particular cutter and metal, as given in Table 13-2
- r/min = the number of revolutions per minute of the milling cutter

EXAMPLE: Find the feed in inches per minute using a 4 in. diameter 12-tooth helical cutter to cut machine steel (CS 80). It would first be necessary to calculate the proper r/min for the cutter.

$$r/min = \frac{CS \times 4}{D}$$
$$= \frac{80 \times 4}{4}$$
$$= 80$$

Feed (in./min) = $N \times$ c.p.t. $\times r/min$
$= 12 \times 0.010 \times 80$
$= 9.6$ or 10 in./min

Metric Calculations

The formula used to find the work feed in millimetres per minute is the same as the formula used to find the feed in inches per minute, except that mm/min is substituted for in./min:

mm/min = $N \times$ chip per tooth $\times r/min$

EXAMPLE: Calculate the feed in millimetres per minute for a 75 mm diameter, six-tooth helical milling cutter when machining a cast iron workpiece (CS 60).

First calculate the r/min of the cutter.
$$r/min = \frac{CS \times 320}{\text{Diameter of cutter}}$$
$$= \frac{60 \times 320}{75}$$
$$= 256$$

Feed (mm/min) = $N \times$ c.p.t. $\times r/min$
$= 6 \times 0.18 \times 256$
$= 276.4$
$= 276$ mm/min

GRADUATED COLLARS

Figure 13-18 shows the three important graduated collars required for most operations done on a milling machine. Each feed screw is fitted with a collar graduated in either hundredths of a millimetre or thousandths of an inch. This collar is free to revolve on a sleeve, but can be locked by a thumb screw to a position relative to the zero or index line as in Figure 13-18. The graduated collars represent the amount of movement in either hundredths of a millimetre or thousandths of an inch of the table, knee, or saddle. The number of graduations on the collar is directly related to the pitch in millimetres or the number of threads per inch on the feed screw. For example, if the crossfeed screw on the saddle has five threads per inch, one revolution of the screw advances the saddle 0.200 in. A crossfeed screw with a pitch of 5 mm would advance the saddle 5 mm in a complete revolution.

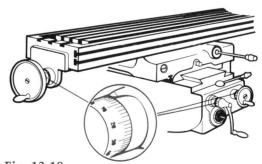

Fig. 13-18
The graduated collars permit accurate settings for horizontal, vertical, and longitudinal depths of cuts.

Courtesy Kostel Enterprises Ltd.

Fig. 13-19
Swivel the vise until the indicator line is aligned with the zero mark.

ALIGNING A VISE

Whenever a workpiece is required to be machined to a layout or to have cuts square or parallel to an edge, the vise must first be aligned. The vise may be aligned by the following methods.
1. The simplest but least accurate method is to align the lines on the vise and the swivel base (Fig. 13-19).
2. The vise may be aligned at right angles to the table travel by placing one edge of a steel square against the column face and the other edge against the solid jaw of the vise (Fig. 13-20).
   When the full width of the vise jaw bears against the face of the square, the locknuts on the vise should be tightened.

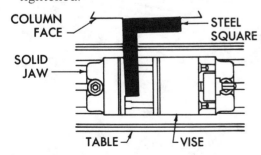

Fig. 13-20
Aligning the solid jaw of the vise with a square held against the column face.

Courtesy Kostel Enterprises Ltd.

Fig. 13-21
When accuracy is required, the indicator is mounted on the milling machine arbor.

Courtesy Kostel Enterprises Ltd.

Fig. 13-22
A parallel gripped in the vise provides a bearing surface for the indicator when aligning a vise.

3. If greater accuracy is required, the vise should be aligned with a dial indicator mounted on the arbor (Fig. 13-21).
   (a) Mount the indicator on the arbor.
   (b) Mount a parallel in the vise (Fig. 13-22).
   (c) Move the parallel up to the indicator button and adjust until there is a reading of approximately 0.50 mm (0.020 in.).
   (d) Move the table until the indicator bears on the other end of the parallel.

(e) Loosen the locknuts and adjust the vise one-half the difference in the indicator readings.

(f) Repeat steps (d) and (e) until the reading is the same at both ends of the parallel.

If the vise must be set at 90° to the table or the column, the same procedure may be followed and the parallel moved across the indicator with the crossfeed handle.

SETTING THE CUTTER TO TO THE WORK SURFACE

Before setting a depth of cut, the operator should check that the work and the cutter are properly mounted and that the cutter is revolving in the proper direction.

To Set the Cutter to the Work Surface

1. Raise the work to within 6 mm (1/4 in.) of the cutter, and directly under it.
2. Place a *long piece* of thin paper on the surface of the work and start the cutter rotating as in Fig. 13-23. Grip the paper *loosely* between the thumb and forefinger.

NOTE: Have the paper long enough to prevent the fingers from coming in contact with the revolving cutter.

3. With the left hand on the vertical feed screw handle, move the work up slowly until the cutter grips the paper.
4. Stop the spindle. Move the knee up 0.05 mm (or 0.002 in.) for paper thickness and set the graduated collar to zero.
5. Move the work clear of the cutter and raise the table to the desired depth of cut.

If the knee is moved up beyond the desired amount, turn the handle backward one half turn and come up to the required line. This will take up the backlash in the thread movement.

This method can also be used when setting the edge of a cutter to the side of a piece of work. In this case, the paper will be placed between the side of the cutter and the side of the workpiece.

Milling a Flat Surface

A milling machine vise can be used to hold square and rectangular work for machining.

1. Align the vise to the column face of the milling machine.
2. Remove all burrs from the workpiece with a file (Fig. 13-24).

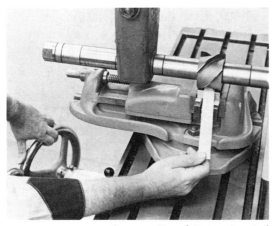

Courtesy Kostel Enterprises Ltd.

Fig. 13-23
Use a long strip of paper between the cutter and the workpiece when setting the cutter to a work surface.

Courtesy Kostel Enterprises Ltd.

Fig. 13-24
Remove all burrs from the work with a file.

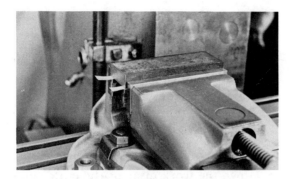

Fig. 13-25
Place short paper feelers between each corner of the work and the parallels.

3. Set the work in the vise, using parallels and paper feelers to make sure that the work is seated properly on the parallels (Fig. 13-25).
4. *Tighten the vise securely* on the workpiece.
5. Tap the work lightly until the paper feeler is tight between the work and the parallels (Fig. 13-26).
6. Mount a plain helical cutter wider than the work to be machined.
   It should be large enough in diameter to allow clearance between the work and the arbor (Fig. 13-27).

NOTE: Be sure that the teeth are pointing in the proper direction for the spindle rotation.

7. Set the proper spindle speed and material (Table 13-1).
8. Set the feed to approximately 0.07 to 0.12 mm (or to 0.003 to 0.005 in.) chip per tooth.
9. Start the cutter and raise the work, using a paper feeler between the cutter and the work (Fig. 13-23).
10. Stop the spindle when the cutter just grips the paper.
11. Set the graduated collar on the elevating screw to zero, allowing for the paper thickness.
12. Move the work clear of the cutter, and set the depth of cut using the graduated collar.
13. For roughing cuts, use a depth of not less than 3 mm (or 1/8 in.) and 0.25 to 0.65 mm (or 0.010 to 0.025 in.) for finish cuts.
14. Set the table dogs for the length of cut.
15. Engage the longitudinal feed and machine the surface.
16. Set up and cut the remaining sides as required.
    See "Machining a Block Square and Parallel" in the Vertical Mill section, for the proper setup if the sides of the workpiece must be machined square and parallel.

Fig. 13-26
Lightly tap the work down onto the parallels with a soft hammer until all paper feelers are tight.

Fig. 13-27
Check for clearance between the arbor support and the work.

Courtesy Kostel Enterprises Ltd.

Fig. 13-28
Side milling a vertical surface.

SIDE MILLING

Side milling is often used to machine a vertical surface on the sides or the ends of a workpiece (Fig. 13-28).

1. Remove all burrs from the workpiece with a file (Fig. 13-24).
2. Set up the work securely in a vise and on parallels.
   Be sure that the surface projects about 12 mm (1/2 in.) beyond the edge of the vise and the parallels to prevent the cutter, vise, or parallels from being damaged.
3. Mount a side milling cutter as close to the spindle bearing as possible to provide maximum rigidity when milling (Fig. 13-29).
   Be sure that the cutter teeth are pointing in the proper direction for the cutter rotation.
4. Set the proper speed and feed for the cutter being used.
5. Start the machine and move the table until the top corner of the work just touches the revolving cutter.
6. Set the crossfeed graduated collar to zero.
7. With the table handwheel, move the work clear of the cutter.
8. Set the required depth of cut with the crossfeed handle.

Courtesy Kostel Enterprises Ltd.

Fig. 13-29
Mount the milling cutter as close as possible to the spindle.
NOTE: The arbor nut is only hand-tight because there is no arbor support mounted.

9. Take the cut across the surface, using the longitudinal automatic feed.

CUTTING SLOTS AND KEYWAYS

Although keyways and slots are generally cut with an end mill, they may be cut on a horizontal mill, using a side milling cutter.

Centring a Cutter to Mill a Slot

1. Locate the cutter as close to the centre of the work as possible.

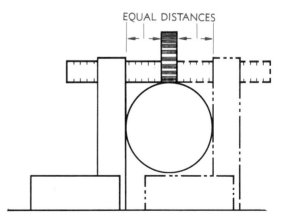

Fig. 13-30
Locating the cutter over the centre of a round shaft.

SIDE MILLING 213

2. Using a steel square and rule, or an inside caliper, adjust the work by using the crossfeed screw until the distance from the cutter to the blade of the square is exactly the same on both sides (Fig. 13-30).
3. Lock the saddle to prevent any movement during the cut.
4. Move the work clear of the cutter and set the depth of cut.
5. Cut the slot to length, using the same procedure in milling a flat surface.

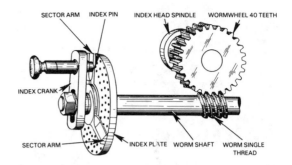

Fig. 13-32
The construction and parts of an indexing mechanism for simple indexing.

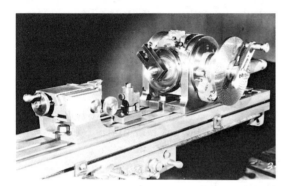

Courtesy Cincinnati Milacron Inc.

Fig. 13-31
An indexing or dividing head set.

THE INDEX OR DIVIDING HEAD

The index head (Fig. 13-31) is a device used to divide the circumference of a piece of work into any number of equal parts, and to hold the work in the required position while the cuts are being made. The most essential parts of the index head are the worm and worm wheel, index crank, index plates, and sector arms (Fig. 13-32).

The index head consists of a 40-tooth worm wheel fastened to the *index head spindle* engaging with a *single-threaded worm* attached to the *index crank*. Since there are 40 teeth in the worm wheel, one complete turn of the index crank will cause the spindle to rotate 1/40 of a turn.

Therefore, 40 turns of the index crank revolve the spindle 1 complete turn, thus making a ratio of 40 to 1.

Work may be indexed by either *simple* or *direct* indexing. The formula for calculating the number of turns for simple indexing is:

$$\frac{40}{\text{Number of divisions to be cut}} = \text{Number of turns of the index crank}$$

EXAMPLE: For 4 divisions,

$$\frac{40}{4} = 10 \text{ full turns of the index crank.}$$

If it is required to cut a reamer with 8 equally spaced flutes, the indexing would be:

$$\frac{40}{8} = 5 \text{ full turns.}$$

For six divisions it would be:

$$\frac{40}{6} = 6\text{-}2/3 \text{ turns.}$$

The six turns are easily made, but what of the 2/3 of a turn? This 2/3 of a turn involves the use of the index plate and sector arms.

The *index plate* is a circular plate having a series of hole circles. Each hole circle contains a different number of equally spaced holes into which the index crank pin can be engaged. The index plate and the crank are used along with the *sector arms*, which eliminates the need for counting a given number of holes each time the work is indexed or turned.

TABLE 13-3 INDEX PLATE HOLE CIRCLES			
Brown & Sharpe		**Cincinnati Standard Plate**	
Plate 1	15-16-17-18-19-20	One Side	24-25-28-30-34-37-38-39-41-42-43
Plate 2	21-23-27-29-31-33		
Plate 3	37-39-41-43-47-49	Other Side	46-47-49-51-53-54-57-58-59-62-66

To get 2/3 of a turn, choose any circle with a number of holes that is evenly divisible by 3, such as 24 (Table 13-3); then take 2/3 of 24 = 16 holes on a 24-hole circle. Thus, for the 6 divisions, we have 6 full turns plus 16 holes on a 24-hole circle.

To Set the Sector Arms

1. To carry out the example using the six turns and 16 holes on a 24-hole circle, place the bevelled edge of a sector arm against the index crank pin, usually in the top hole, then count 16 holes on the 24-hole circle, *not including the hole in which the index crank pin is engaged* (Fig. 13-33A).
2. Move the other sector arm just beyond the sixteenth hole, and tighten the set screws.
3. After the first cut has been made and the cutter returned for the next cut, withdraw the index crank pin, make six full turns *clockwise*, plus the 16 holes between the sector arms.
4. Stop between the last two holes, gently tap the index crank, and allow the pin to snap into place.
5. Move the sector arms around against the pin, ready for the next division (Fig. 13-33B).

NOTE: If the pin is turned past the required hole, an error will appear in the spacing unless it is turned back at least 1/2 turn to eliminate backlash and then brought back to the proper hole.

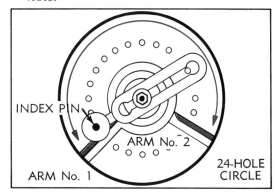

Fig. 13-33B
The position of the sector arms after indexing 6-2/3 turns.

To Mill a Hexagon

A hexagon, 40 mm (or 1-5/8 in.) across flats, is to be milled on a 50 mm (or 2 in.) diameter shaft (Fig. 13-34). To cut the hexagon, check the alignment of the index centres and then mount the work. The indexing for 6 divisions $= \dfrac{40}{6}$ or 6-2/3 turns.

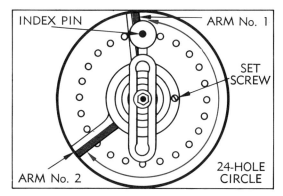

Fig. 13-33A
The sector arms set for 16 holes on a 24-hole circle (2/3 turn).

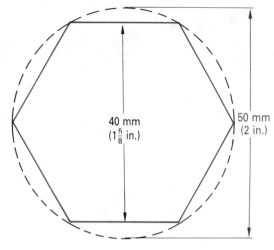

Fig. 13-34
A 40 mm hexagon to be cut on a 50 mm diameter shaft.

The following procedure could be used for milling any hexagon — only the measurements would be changed.
1. Select a hole circle that the denominator 3 will divide into, such as 24.
2. Set the sector arms to 16 holes on a 24-hole circle (Fig. 13-33A).
   Do not count the hole the pin is in.
3. From the hole the pin is in, count 16 holes and adjust the other sector arm to just beyond the sixteenth hole (Fig. 13-33B).

Courtesy Kostel Enterprises Ltd.

Fig. 13-35
Milling the second flat of the hexagon.

4. Start the machine and set the cutter to the top of the work by using a paper feeler (Fig. 13-23).
5. Set the graduated dial at zero.
6. Calculate the depth of cut (Fig. 13-34). 50 mm − 40 mm = 10 mm (or 2.000 in. − 1.625 in. = 0.375 in.). As only one-half of the material is to be removed from each side, the depth of
   cut = $\frac{10}{2}$ = 5 mm (or $\frac{0.375}{2}$ = 0.1875 in.).
7. Move the work clear of cutter.
8. Raise the table 5 mm (or 0.1875 in.).
9. Lock the knee clamp.
10. Mill the first flat.
11. Index 20 turns (3 × 6-2/3) and mill the opposite side (Fig. 13-35).

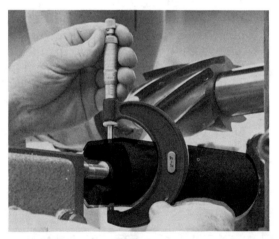

Courtesy Kostel Enterprises Ltd.

Fig. 13-36
Measuring the distance across the flats.

12. With a micrometer, check the distance across the flats and adjust and recut if necessary (Fig. 13-36).
13. Continue indexing 6-2/3 turns until all six sides are cut.

PLAIN OR DIRECT INDEXING

Direct indexing is the simplest form of indexing, but it can only be used for milling divisions that are divisible into 24, 30,

TABLE 13-4 DIVISIONS FOR WHICH PLATE CAN BE USED													
Plate Hole Circles or Slots													
24	2	3	4	—	6	8	—	—	12	—	—	24	—	—
30	2	3	—	5	6	—	—	10	—	15	—	—	30	—
36	2	3	4	—	6	—	9	—	12	—	18	—	—	36

or 36. The common divisions that can be obtained are listed in Table 13-4.

To Obtain Direct Indexing

1. Disengage the worm from the wheel by turning the handle counterclockwise. This disengages the spindle from the index crank.
2. Mount the proper direct indexing plate with the holes or slots facing the plunger pin.
3. Mount the workpiece in the dividing head chuck or between centres.
4. Lock the spindle and cut the first surface or groove.
5. Disengage the plunger pin from the indexing plate and unlock the spindle (Fig. 13-37).
6. Turn the indexing plate the proper number of holes or slots.
7. Engage the plunger pin and lock the spindle.
8. Cut the remaining surfaces or grooves in the same manner.

EXAMPLE: What direct or plain indexing is required to mill four flats on a round shaft?

Indexing $= \dfrac{24}{4}$ 6 holes or slots in 24-hole circle for each flat milled

OR

$\dfrac{36}{4} =$ 9 holes or slots in 36-hole circle for each flat milled.

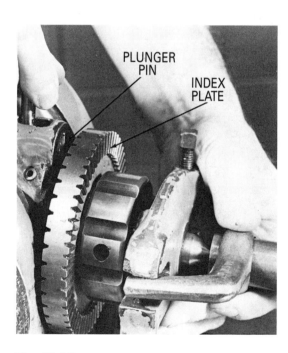

Fig. 13-37
The direct indexing plate, attached to the dividing head spindle, is used for indexing a limited number of divisions.

VERTICAL MILLING MACHINES

A vertical milling machine (Fig. 13-38) has basically the same parts as a horizontal plain milling machine. Instead of the cutter fitting into a horizontal spindle, it fits into a vertical spindle. On most machines the head can be swivelled 90° to either side of the centre line for the cutting of angular surfaces. The vertical milling machine is especially useful for face and end milling operations.

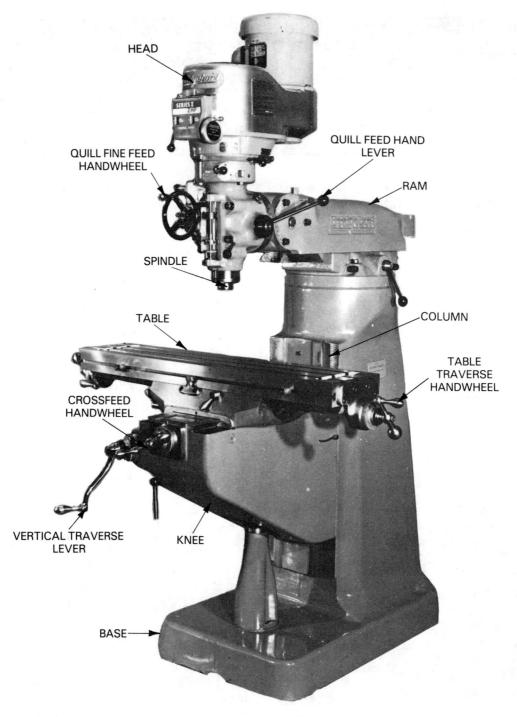

Courtesy Bridgeport Machines
Division of Textron Inc.

Fig. 13-38
A vertical milling machine is useful for face and end milling operations.

VERTICAL MILL PARTS

The *base* (Fig. 13-38) is made of ribbed cast iron. It may contain a coolant reservoir.

The *column* is cast in one piece with the base. The machined face of the column provides the ways for the vertical movement of the knee. The upper part of the column is machined to receive a *turret* on which the *overarm* is mounted.

The *overarm* may be round or of the more common *dovetailed ram* type. It may be adjusted toward or away from the column and swivelled to increase the capacity of the machine.

The *head* is attached to the end of the ram (or overarm) (Fig. 13-38). On universal type machines, the head may be swivelled in two planes. The motor, which provides the drive to the *spindle*, is mounted on top of the head. Spindle speed changes may be made by means of gears and V-belts or variable speed pulleys on some models. The spindle mounted in the *quill* may be fed by means of the *quill feed hand lever*, the *quill fine feed handwheel*, or by automatic power feed.

The *knee* moves up and down the face of the column and supports the *saddle* and *table*. On most machines, all table movements are controlled manually. Automatic feed control units for any table movements are usually added as accessories.

MILLING CUTTERS AND COLLETS

Most machining on the vertical mill is done with either an end mill, a shell end mill, or a flycutter.

Courtesy Weldon Tool Co.

A - *Two-fluted end mill*

Courtesy Butterfield Division, Union Twist Drill Co.

B - *Four-fluted end mill*

Courtesy Cleveland Twist Drill (Canada) Ltd.

C - *Shell end mill*

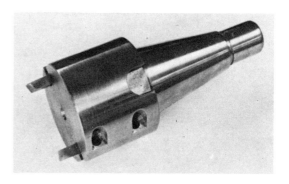

Courtesy Kostel Enterprises Ltd.

D - *Flycutter*

Fig. 13-39
Types of vertical milling cutters.

End mills have cutting teeth on the end as well as on the periphery and are fitted to the spindle by a suitable adaptor. They may be of two types, the *solid end mill* (Fig. 13-39A, B) or the *shell end mill* (Fig. 13-39C), which is fitted to a separate shank.

The solid end mills may have two or more flutes. Two-fluted end mills (Fig. 13-39A) have different-length lips on the end and may be used for drilling shallow holes. The four-fluted type (Fig. 13-39B) requires a starting hole before milling a slot in the centre of a workpiece.

Larger work surfaces may be machined by means of a *flycutter* (Fig. 13-39D), which holds two or more single-pointed cutting tools. Flycutters provide an economical means of machining work surfaces.

TYPES OF COLLETS

There are two main types of collets — the spring type and the solid type. Both are driven by a key in the spindle bore and a keyway on the outside of the collet.

The *spring collet* (Fig. 13-40A) holds and drives the cutter by means of friction between the collet and cutter. On heavy cuts the cutter may move up in the collet if it is not tightened securely.

The *solid collets* (Fig. 13-40B) are more rigid and hold the cutter securely. The solid collets may be driven by a key in the spindle and a keyway in the collet, or by two drive keys on the spindle. The cutter is driven and prevented from turning by one or two setscrews in the collet, which bear against flats on the cutter shank.

VERTICAL MILLING MACHINE OPERATIONS

MOUNTING AND REMOVING CUTTERS

The vertical milling machine permits the use of a wide variety of cutting tools. These tools may be held in the spindle by means of a spring collet or an adaptor which is held in the spindle by means of a draw-in bar (Fig. 13-41).

Mounting a Cutter in a Spring Collet

1. Shut off the electric power to the machine.

Courtesy Kostel Enterprises Ltd.

A - Spring

Courtesy Weldon Tool Co.

Courtesy Weldon Tool Co.

B - Solid

Fig. 13-40
Types of vertical milling collets.

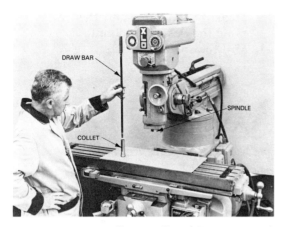

Courtesy Kostel Enterprises Ltd.

Fig. 13-41
Mounting a spring collet in a vertical mill spindle.

Courtesy Kostel Enterprises Ltd.

Fig. 13-42
Tightening the draw-in bar to secure the collet and cutter in the spindle.

2. Place the proper cutter, collet, and wrench on a piece of masonite on the machine table.
3. Clean the taper in the spindle.
4. Place the draw-in bar into the hole in the top of the spindle.
5. Clean the taper and keyway on the collet.
6. Insert the collet into the bottom of the spindle, press up, and turn it until the keyway aligns with the key in the spindle.
7. Hold the collet up with one hand and with the other, thread the draw-in bar into the collet for about four turns.
8. Hold the cutting tool with a rag and insert it into the collet for the full length of the shank.
9. Tighten the draw-in bar into the collet (clockwise) by hand.
10. Hold the spindle-brake lever and tighten the draw-in bar securely with a wrench (Fig. 13-42).

Removing a Cutter from a Collet
1. Shut off the electric power to the machine.
2. Place a piece of masonite on the machine table to hold the necessary tools.
3. Pull on the spindle brake lever and loosen the draw-in bar with a wrench (counterclockwise).
4. Loosen the draw-in bar by hand, only about three turns.
   NOTE: Do not unscrew the draw-in bar from the collet.
5. With a soft-faced hammer, strike down sharply on the head of the draw-in bar to break the taper contact between the collet and spindle.
6. With a cloth, remove the cutter from the collet.
7. Clean the cutter and replace it in its proper storage place.

Mounting a Cutter in a Solid Collet
1. Shut off the electric power to the machine.
2. Place the cutter, collet, and necessary tools on a piece of masonite on the machine table.
3. Slide the draw-in bar through the top hole in the spindle.
4. Clean the spindle taper and the taper on the collet.
5. Align the keyway or slots of the collet with the keyway or drive keys in the spindle, and insert the collet into the spindle.

VERTICAL MILLING MACHINE OPERATIONS

6. Hold the collet up in the spindle and thread the draw-in bar clockwise with the other hand.
7. Pull on the brake lever and tighten the draw-in bar securely with a wrench.
8. Insert the end mill into the collet until the flat(s) align with the setscrew(s) of the collet.
   To remove an end mill from a solid collet reverse the procedure.

To Machine a Flat Surface

1. Check that the vertical head is at right angles to the table in both directions, so that a flat surface is produced.
2. Mount a suitable flycutter in the machine spindle (Fig. 13-43).

Courtesy Kostel Enterprises Ltd.

Fig. 13-43
Machining a flat surface with a flycutter.

3. Set the machine speed for the size of cutter and material being milled.
4. Remove all burrs from the workpiece.
5. Clean the work and the vise.
6. Mount the work in the vise, on suitable parallels and with paper feelers under each corner (Fig. 13-25).
7. Start the machine, and raise the table until the cutter just touches the surface of the work.
8. Set the vertical feed graduated collar to zero.
9. Raise the table 0.25 mm or 0.010 in. and take a trial cut approximately 12 mm (1/2 in.) long.

10. Measure the work and then raise the table the desired amount.
11. Mill the surface to size.

Courtesy Kostel Enterprises Ltd.

Fig. 13-44
The sequence for machining the sides of a rectangular workpiece square and parallel.

MACHINING A BLOCK SQUARE AND PARALLEL

In order to machine the four sides of a piece of work so that its sides are square and parallel, it is important that each side be machined in a definite order (Fig. 13-44). It is very important that dirt and burrs be removed from the work, vise, and parallels, since dirt and burrs can cause inaccurate work.

Courtesy Kostel Enterprises Ltd.

Fig. 13-45
Side #1 or the surface with the largest area should be facing up.

Machining Side #1

1. Check that the vertical head is at right angles to the table in both directions.
2. Remove all burrs from the workpiece.
3. Clean the work and the vise.
4. Mount the work in the centre of the vise, on parallels, with its largest side (#1) up and use paper feelers under each corner (Fig. 13-45).
5. Mount a flycutter in the milling machine spindle.
6. Set the machine for the proper speed for the size of the cutter and the material to be machined. (See Table 13-1.)

Courtesy Kostel Enterprises Ltd.

Fig. 13-47
A clean-up cut only is required on side #1.

Courtesy Kostel Enterprises Ltd.

Fig. 13-46
Raise the table until the cutter just touches the end of the work.

7. Start the machine, and raise the table until the cutter just touches near the right-hand end of the side #1 (Fig. 13-46).
8. Move the work clear of the cutter.
9. Raise the table about 1.5 mm (or 0.060 in.) and machine side #1, using a steady feed rate (Fig. 13-47).
10. Take the work out of the vise and remove all burrs from the edges with a file.

Machining Side #2

11. Clean the vise, work, and parallels thoroughly.
12. Place the work on parallels, if necessary, with side #1 against the solid jaw and side #2 up (Fig. 13-48).
13. Place a round bar between side #4 and the movable jaw.
    The round bar must be in the centre of the amount of work being held inside the vise jaws.
14. Tighten the vise securely and tap the work down until the papers feelers are tight.
15. Follow steps 7 to 10 and machine side #2.

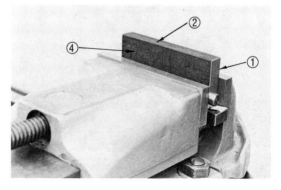

Courtesy Kostel Enterprises Ltd.

Fig. 13-48
The finished side (#1) is placed against the solid jaw in order to machine side #2 square.

VERTICAL MILLING MACHINE OPERATIONS

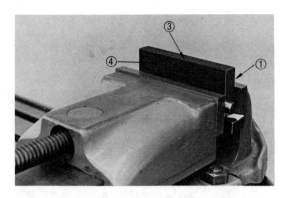

Courtesy Kostel Enterprises Ltd.

Fig. 13-49
The setup required to machine side #3. The width should be to size after this operation.

Machining Side #3

16. Clean the vise, work, and parallels thoroughly.
17. Place side #1 against the solid vise jaw with side #2 resting on parallels if necessary (Fig. 13-49).

    Move the parallel to the left so that about 6 mm (1/4 in.) of the work extends beyond the end of the parallel. This permits the work to be measured while in the vise.

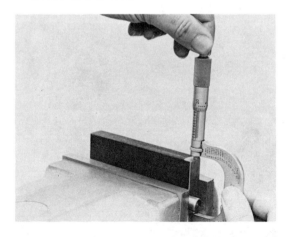

Courtesy Kostel Enterprises Ltd.

Fig. 13-50
Measuring the work after the trial cut.

18. Place a round bar between side #4 and the movable jaw.

    The round bar must be in the centre of the amount of work being held inside the vise jaws.

19. Tighten the vise securely and tap the work down until the paper feelers are tight.
20. Start the machine and raise the table until the cutter just touches near the right-hand end of side #3.
21. Move the work clear of the cutter and raise the table about 0.25 mm or 0.010 in.
22. Take a trial cut about 6 mm (1/4 in.) long, stop the machine, move the work clear of the cutter, and measure the width of the work (Fig. 13-50).
23. Raise the table the required amount and mill side #3 to the correct width.
24. Remove the work and file off all burrs.

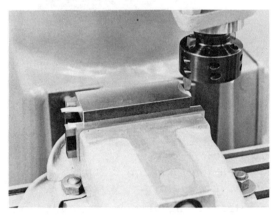

Courtesy Kostel Enterprises Ltd.

Fig. 13-51
Setup for machining side #4.

Machining Side #4

25. Clean the vise, work, and parallels thoroughly.
26. Place side #1 down on the parallels with side #4 up and tighten the vise securely (Fig. 13-51).

    With three finished surfaces, the round bar is not required when milling side #4.

27. Tap the work down until the paper feelers are tight.
28. Follow steps 20 to 23 and machine side #4 to the correct thickness.

MACHINING THE ENDS OF A BLOCK OR WORKPIECE

The ends of the workpiece may be machined by two methods, depending on the length and shape of the workpiece. Short pieces (no more than about 90 mm or 3-1/2 in. long) may be held upright in the centre of the vise and machined with a flycutter or shell end mill. Long workpieces must be gripped in the vise, which is set parallel to the table travel and machined with an end mill.

Flycutter or Shell End Mill Method

1. Remove the burrs from the workpiece.
2. Clean the vise and the work.
3. Set the work in the centre of the jaws with one end on the base of the vise and tighten the vise lightly.
4. Tap the work down on the base and square it with the solid vise jaw as shown in Fig. 13-52.
5. Tighten the vise securely.
6. Centre the workpiece with the cutter and take a cleanup cut using the cross-feed handle.

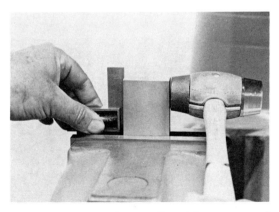

Courtesy Kostel Enterprises Ltd.

Fig. 13-52
Squaring a short workpiece before machining the end.

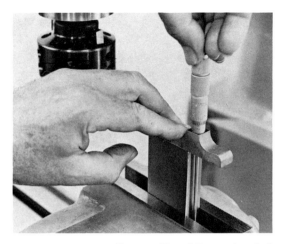

Courtesy Kostel Enterprises Ltd.

Fig. 13-53
Measuring the length of the workpiece after a clean-up cut.

7. Remove the work from the vise, remove the burrs, and check for squareness.
8. Clean the vise and the work thoroughly, and place the machined end on paper feelers on the base of the vise.
9. Tighten the vise securely and tap the work down onto the feelers.
10. Take a cleanup cut off the end.
11. Remove the burrs and measure the height of the work (Fig. 13-53).
12. Raise the table the required amount and machine the work to length.

End Mill Method

1. Set the toolhead square (90°) in both directions.
2. Align the vise parallel to the table travel.
3. Remove the burrs from the workpiece.
4. Clean the workpiece and the vise throughly.
5. Mount the work in the vise on suitable parallels and on paper feelers (Fig. 13-54).

The work should extend past the vise jaws by at least 6 mm (1/4 in.).

Courtesy Kostel Enterprises Ltd.

Fig. 13-54
The workpiece should project about 6 mm (1/4 in.).

Courtesy Kostel Enterprises Ltd.

Fig. 13-56
A long paper feeler is used to touch the cutter up to the work.

6. Tighten the vise *securely*; then tap the work onto the parallels until all the paper feelers are tight.
7. Mount a suitable end mill in the spindle (Fig. 13-55).
8. Set the proper *r/min* and check that the cutter is revolving in a clockwise direction.
9. Centre the cutter with the work and move the table until the *revolving* end mill just cuts along paper feeler between the work and the cutter (Fig. 13-56).
10. Move the work clear of the cutter, using the crossfeed handle.
11. Set a 0.80 to 1.50 mm (0.030 to 0.060 in.) depth of cut with the longitudinal table feed handle (Fig. 13-57) and tighten the table traverse lock.
12. Take the cut across the end of the work, using the crossfeed handle.
13. Remove the work, remove the burrs, and check for squareness.
14. Repeat the procedure on the other end and machine the work to length.

Courtesy Kostel Enterprises Ltd.

Fig. 13-55
Mount the end mill and check the direction of spindle rotation.

Courtesy Kostel Enterprises Ltd.

Fig. 13-57
The depth of cut is set with the longitudinal feed handle.

PRODUCING AND FINISHING HOLES

Drilling, tapping, counterboring, countersinking, and boring operations can be done accurately on a vertical mill. The spindle is adaptable for the use of the tools for these operations, and the feed screw graduated collars provide accurate locating of the holes to be machined.

NOTE: The vertical head must be at right angles (90°) to the table before doing any of these operations.

To Drill on a Vertical Mill

1. Mount the work in a vise or clamp it to the table.
   Work must be supported by parallels which are positioned so that they will not interfere with the drill.
2. Mount a drill chuck in the spindle.
3. Mount a centre finder in the drill chuck.
4. Adjust the table until the centre punch mark of the hole to be drilled is in line with the tip of the rotating centre finder.
5. Tighten the table and saddle clamps.
6. Stop the machine and remove the centre finder.
7. Place a centre drill in the chuck and slightly drill the centre hole.
8. Mount the proper size drill in the drill chuck.
9. Set the machine to the proper speed and feed.
10. Engage the quill feed and drill the hole, using cutting fluid.

To Ream on a Vertical Mill

After the hole has been drilled, the table should not be moved, in order that the reamer and the drilled hole will be aligned.

1. Mount the reamer in the spindle or drill chuck.
2. Set the speed and feed for reaming (approximately 1/4 of the drilling speed).
3. Apply cutting fluid as the reamer is fed into the hole.
4. Stop the machine.
5. Remove the reamer from the hole. (Do not turn reamer backwards.)

To Mill Slots and Keyways

Keyways, keyseats, and slots can be easily and quickly cut in a vertical milling machine by using an end milling cutter.

1. Layout the keyseat and the end of the shaft as in Fig. 13-58.

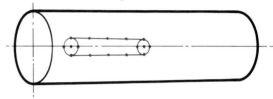

Fig. 13-58
Layout of a keyseat on a shaft.

2. Place the work in the vise, or in V blocks for longer work.
3. Using a square, align the end layout line, which will set the keyseat to proper position on the top of the shaft (Fig. 13-59).

Courtesy Kostel Enterprises Ltd.

Fig. 13-59
Setting-up the work to position the keyseat for milling.

4. Fasten the vise securely.
5. Mount a two or three-fluted end mill. The diameter of the end mill must be the width of the keyseat.
6. Centre the workpiece by touching the revolving end mill to a piece of paper held against the shaft (Fig. 13-60).

Fig. 13-60
Setting the cutter to the side of the work.

7. Set the crossfeed graduated collar to zero.
8. Raise the table until the revolving end mill clears the work.
9. Move the table over a distance of half the diameter of the work plus half the diameter of the cutter.
10. Adjust the table until the end mill is in line with one end of the keyseat.
11. Feed the table up until the end mill cuts to its full diameter on the shaft.
12. Note the reading on the graduated collar on the vertical traverse screw shaft.

Fig. 13-61
The keyseat machined to the proper length.

13. Raise the table until the depth of cut is one-half the thickness of the key.
14. Lock the knee clamp and machine the keyseat to the proper length (Fig. 13-61).

TEST YOUR KNOWLEDGE
Milling Machines and Parts
1. List six types of work that can be produced on a vertical milling machine.
2. What is the difference between a plain horizontal and a universal milling machine?
3. What hand controls are used to move the table
   (a) lengthwise?
   (b) in or out?
   (c) up or down?
4. What precaution should be observed before changing the spindle speed?

Milling Machine Safety
5. List three safety precautions you consider most important, and explain why.
6. Name two materials used to make milling cutters.
7. Name three types of milling cutters and state the purpose of each.

Mounting and Removing an Arbor
8. How is the cutter prevented from turning on the arbor?
9. Why is it necessary to clean the spindle and arbor tapers before mounting an arbor?
10. What is the purpose of the draw-in bar?
11. Explain in point form how to remove an arbor.

Mounting and Removing a Milling Cutter
12. What is the purpose of the masonite when changing a cutter?
13. What should be checked before starting to mill a surface?
14. Why is it important that the arbor support be in place before using a wrench on the arbor nut?

Work-Holding Devices

15. Name four commonly used holding devices.
16. What holding device can be used when many similar pieces are machined?

Cutting Speeds

17. Define cutting speed.
18. Name five important factors to consider when setting the *r/min* to machine a surface.
19. What *r/min* is required to mill a piece of cast iron (*CS* 60) using a 4 in. diameter high-speed cutter?
20. Calculate the *r/min* required to mill a piece of tool steel (*CS* 55 m/min) using a 75 mm carbide tipped cutter.

Milling Feeds

21. Define feed.
22. How is the milling feed determined?
23. What feed is recommended for milling machine steel (*CS* 100) using a 6 in. diameter 24-tooth side milling cutter?
24. Calculate the feed for milling a piece of bronze (*CS* 30 m/min) using a 100 mm, 6-tooth high-speed plain helical milling cutter.

Graduated Collars

25. What feed screws are equipped with graduated collars?
26. What is the value of each division on the graduated collar?
27. If the sides of the work do not have to be square, how may the vise be aligned?
28. Name and briefly describe two other methods of aligning a vise on a milling machine.

Setting the Cutter to the Work Surface

29. It is required to remove 3/16 in. from a surface. Explain how a graduated collar can be used to do this to accuracy of 0.001 in.
30. Explain how a graduated collar may be used to accurately remove 3.18 mm from a workpiece.
31. Explain briefly how to set a cutter to the work surface before setting a depth of cut.

Milling a Flat Surface

32. State two methods by which a vise may be aligned.
33. What is the purpose of using paper feelers when setting up work on parallels?
34. By means of suitable sketches show how a rectangular piece of steel is set up to machine the four sides square and parallel.

Side Milling

35. What is the purpose of side milling?
36. What precaution should be observed when setting up work for side milling?
37. Why should the side milling cutter be mounted as close to the spindle as possible?
38. What graduated collar is used to set the depth of cut when side milling?

Centring a Cutter

39. List in point form how to centre a 1/4 in. keyway cutter to a 1 in. shaft.
40. List the steps required to centre a 6 mm keyway cutter to a 30 mm shaft.

The Index or Dividing Head

41. What is the purpose of the dividing head?
42. What two types of indexing may be done on the dividing head?
43. What is the formula for calculating simple indexing?
44. Calculate the simple indexing for 24, 27, and 36 divisions using the Brown and Sharpe plate.
45. What procedure should be followed to set the sector arms for 15 holes in the 20-hole circle?
46. What will be the depth of cut required to mill a hexagon 19 mm across the flats from a 30 mm diameter piece of material?
47. After one side of a hexagon has been machined, what side should be milled next? Explain why.
48. How does direct indexing differ from simple indexing?

49. What direct indexing is required to cut
    (a) a hexagon?
    (b) an octagon?

Vertical Milling Machine

50. How does the vertical mill differ from a horizontal mill?
51. State the purpose of the following vertical mill parts:
    (a) column
    (b) overarm
    (c) knee

Milling Cutters and Collets

52. Name three types of cutters used on a vertical mill.
53. Describe an end mill.
54. For what purpose are flycutters used?
55. Name two types of collets used in a vertical mill.

To Machine a Flat Surface

56. When machining a flat surface, why must the vertical head be at right angles (90°) to the table?

Milling a Block Square and Parallel

57. How is the work set up to mill side #1?
58. How is the work set up to mill side #2?
59. Where should the round bar be placed when milling sides #2 and #3?
60. How is the work set up for milling side #4?

Machining the Ends of a Block

61. Why it is important to remove the burrs from a workpiece before mounting it in a vise?
62. What is the purpose of the paper feelers?
63. Briefly describe how the end of a short block may be squared.
64. How is longer work held for squaring the end?

Producing and Finishing Holes

65. Explain briefly how the following operations can be performed:
    (a) drilling
    (b) reaming

To Mill Slots and Keyways

66. Explain how to set up a shaft in a vise to mill a keyseat.
67. Briefly describe how to centre the cutter after the shaft has been set up.

CHAPTER 14
GRINDERS

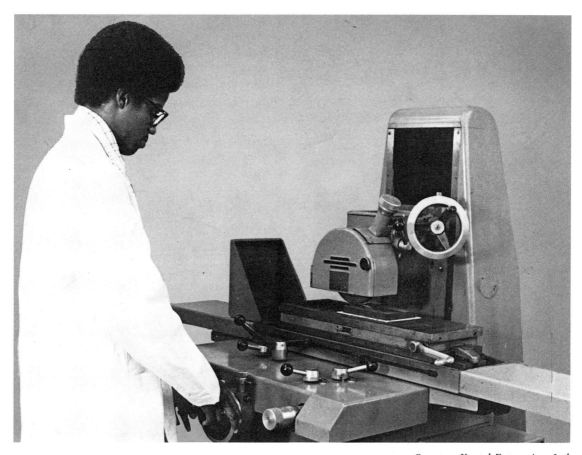

Courtesy Kostel Enterprises Ltd.

Grinding is a metal removal process which uses an abrasive cutting tool to produce a high surface finish and bring the workpiece to an accurate size. In the grinding process, a revolving grinding wheel is brought into contact with the surface of a workpiece. Each abrasive grain on the periphery of the grinding wheel is a cutting tool, and as it contacts the work surface it removes a minute chip of metal. The modern grinding machine is capable of finishing soft or hardened workpieces to tolerances of 0.005 mm (0.0002 in.) or less on high production runs, while at the same time producing a very high surface finish.

There are various types of grinding machines used in the machine tool trade to suit the sizes and shapes of a wide variety of workpieces. *Bench or pedestal grinders* are used for the sharpening of cutting tools and the rough grinding of metals. *Surface grinders* are used to produce flat, angular, or contoured forms on a flat surface. *Cylindrical grinders* are used to grind round workpieces and can produce internal and external diameters which are straight, tapered, or contoured. There are many other special types of grinders used for specific purposes in the machine tool trade.

BENCH OR PEDESTAL GRINDER

The bench or pedestal grinder is used for the sharpening of cutting tools and the rough grinding of metal. Because the work is usually held in the hand, this type of grinding is called *offhand grinding*.

The bench grinder is mounted on a bench, while the pedestal grinder (Fig. 14-1), being a larger machine, is fastened to the floor. Both types consist of an electric motor with a grinding wheel mounted on each end of the spindle. One wheel is usually a coarse-grained wheel for the fast removal of metal, while the other is a fine-grained wheel for finish grinding. The *U-shaped work rests* provide a rest for either the work or the hands while grinding.

NOTE: Always keep the work rests adjusted within 1.5 mm (1/16 in.) of the wheel to prevent work being jammed between the rest and the wheel.

The *safety glass eyeshields* provide eye protection for the operator, but it is good safety practice to wear safety glasses while grinding, even though the machine is equipped with eyeshields.

GRINDING WHEELS

Aluminum oxide and *silicon carbide* are the two types of grinding wheels generally used on bench or pedestal grinders. These manufactured abrasives are superior to natural abrasives such as emery, sandstone, corundum, and quartz because they contain no impurities and are of a more uniform grain structure.

Aluminum oxide (Fig. 14-2) is made in an arc-type electric furnace by charging bauxite, ground coke, and iron borings. Grinding wheels made from aluminum oxide are used to grind high tensile strength materials such as hard or soft carbon and alloy steels, tough bronze, etc.

Silicon carbide (Fig. 14-3) is manufactured in a resistance-type electric furnace by charging silica sand and coke with

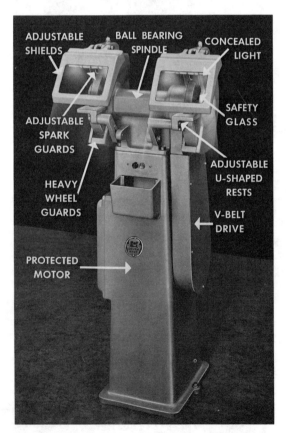

Courtesy South Bend Lathe, Inc.

Fig. 14-1
The main parts of a pedestal grinder.

Courtesy Carborundum Co.

Fig. 14-2
Aluminum oxide is manufactured from bauxite ore, ground coke, and iron borings.

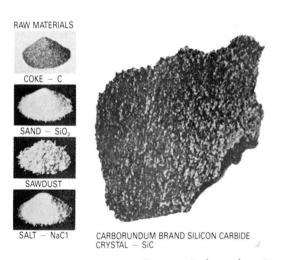

Courtesy Carborundum Co.

Fig. 14-3
Silicon carbide is manufactured from silica sand, coke, sawdust, and salt.

small amounts of sawdust and salt. Grinding wheels made of silicon carbide are used to grind low tensile strength materials such as aluminum, copper, ceramics, cast iron, etc.

GRINDER SAFETY

Because grinding wheels operate at very high speeds and the grinding particles are very fine, it is important to observe the following safety precautions.

1. Always wear approved safety glasses when operating a grinder.
2. Always stand to one side of the wheel when starting a grinder.
   NOTE: Never stand in line with a grinding wheel (Fig. 14-4).

Courtesy Kostel Enterprises Ltd.

Fig. 14-4
Always stand to one side of the wheel when starting a grinder.

3. Allow a new wheel to run for about one minute before using.
   NOTE: If a wheel is going to break, it will break in the first minute.
4. Always use a wheel guard that covers at least one-half of the grinding wheel.
5. Never run a grinding wheel faster than the speed recommended on its blotter.
6. Do not grind on the side of a wheel unless it is designed for this purpose.
7. Never force a grinding wheel by jamming work into it.
8. Be sure that the grinder workrest is within 1.5 mm (1/16 in.) of the grinding wheel face to prevent work from jamming between the wheel and the rest.
9. Always remove burrs produced by grinding with a file (Fig. 14-5).

Courtesy Kostel Enterprises Ltd.

Fig. 14-5
Remove the burrs and sharp edges produced by grinding as soon as possible.

DRESSING AND TRUING A WHEEL

When a grinding wheel is used, several things happen to it:
1. Small metal particles imbed themselves in the wheel, causing it to become loaded or clogged (Fig. 14-6A).
2. The abrasive grains become worn smooth and the wheel loses its cutting action.
3. Grooves become worn in the face of the wheel.

Any one of these reasons requires the immediate *dressing* or *truing* of a wheel. If a loaded wheel is not dressed, it will not cut properly, and it will start to heat, burn, and distort the workpiece. Dressing is the process of reconditioning the wheel to make it cut better (Fig. 14-6B). Truing refers to shaping a wheel to a desired shape and to make its grinding surface run true with its axis. Both truing and dressing may be done at the same time with a mechanical wheel or an abrasive stick dresser.

Courtesy Kostel Enterprises Ltd.

Fig. 14-6A
The face of a grinding wheel after it has been severely loaded.

Courtesy Kostel Enterprises Ltd.

Fig. 14-6B
The same wheel after it has been properly dressed.

Courtesy Norton Co. of Canada, Ltd.

Fig. 14-7
Types of mechanical wheel dressers.

A mechanical wheel dresser (Fig. 14-7), commonly called a star dresser, is usually employed to dress an offhand grinding wheel. It consists of a number of hardened, pointed disks mounted loosely in a handle.

To Dress and True a Wheel

1. Adjust the grinder work rests so that when the lugs of the mechanical dresser are against its edge, the dresser rolls just touch the face of the wheel (Fig. 14-8).
2. Wear an approved pair of safety glasses.
3. Stand to one side of the wheel and then start the grinder.
4. Hold the dresser down firmly with its lugs against the edge of the work rest. *NOTE: DO NOT contact the revolving wheel.*
5. Move the dresser across the face of the wheel in a steady motion for a trial pass.
6. After each pass, tilt the holder up slightly to advance the dresser disks into the wheel.
7. When the wheel is dressed, stop the grinder and adjust the work rests to within 1.5 mm (1/16 in.) of the wheel face.

GRINDING A LATHE TOOLBIT

All lathe toolbits, regardless of shape, must have relief angles and side rake in order for them to cut properly. Since it is impossible to cover the grinding of all types of lathe toolbits, only the general purpose toolbit will be explained in detail.

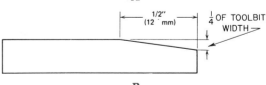

Courtesy Kostel Enterprises Ltd.

Fig. 14-9
Grinding the side-cutting edge and the side-relief angle on a toolbit.

Courtesy Kostel Enterprises Ltd.

Fig. 14-8
Dressing a grinding wheel with a mechanical wheel dresser.

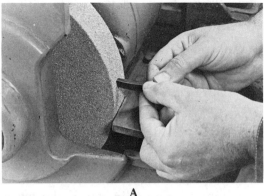

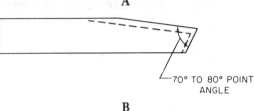

70° TO 80° POINT ANGLE

Courtesy Kostel Enterprises Ltd.

Fig. 14-10
Shaping the point and grinding the end-relief angle.

A toolbit grinding gauge (Fig. 14-12) should be used to check all angles and clearances.

To Grind a Lathe Toolbit

1. Hold the toolbit firmly while supporting the hands on the grinder tool rest.
2. Tilt the bottom of the toolbit in toward the wheel and grind the 10° side-relief angle and form required on the left side of the toolbit (Fig. 14-9).
3. Grind until the side cutting edge is about 1/2 in. (12.0 mm) long and the point is over about 1/4 of the width of the toolbit (Fig. 14-9).
   While grinding, move the toolbit back and forth across the face of the wheel. This helps to grind faster and prevents grooving the wheel.
4. High-speed steel toolbits must be cooled frequently. *Never overheat a toolbit.*
5. Hold the back end of the toolbit lower than the point and grind the 15° end-relief angle on the right side. At the same time the end cutting edge should form an angle of from 70 to 80° with the side cutting edge (Fig. 14-10).
6. Hold the toolbit about 45° to the axis of the wheel (Fig. 14-11), tilt the bottom of the toolbit in, and grind the 14° side rake on the top of the toolbit.
7. Grind the side rake the entire length of the side cutting edge, but do not grind the top of the cutting edge below the top of the toolbit.
8. Grind a slight radius on the point, being sure to keep the same end and side-relief angles.
9. Use an oilstone to hone the point and cutting edge of the toolbit to remove sharp edges and improve its cutting action.

To Sharpen a General-Purpose Toolbit

A general-purpose lathe toolbit, ground to the shape and dimensions shown in Fig. 14-13, can be quickly sharpened by grind-

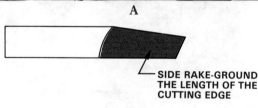

SIDE RAKE-GROUND THE LENGTH OF THE CUTTING EDGE

Courtesy Kostel Enterprises Ltd.

Fig. 14-11
Grinding the side rake on the top of the toolbit.

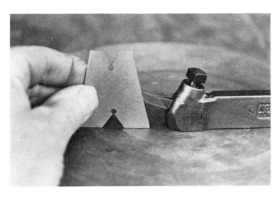

Courtesy Kostel Enterprises Ltd.

Fig. 14-12
Checking the end-relief angle with a toolbit grinding gauge.

ing only the end cutting edge. It is important to maintain the same shape and end-relief angle when sharpening a toolbit.

After the worn portion is removed, grind a slight radius on the point and hone the cutting edge. When the side-cutting edge becomes too short, after repeated sharpenings, regrind the whole toolbit to the original shape and dimensions.

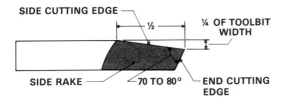

Fig. 14-13
The top view of a general purpose toolbit, showing its shape and dimensions.

DRILL GRINDING

Before using a drill, it is wise to examine its condition. To cut properly and efficiently, a drill should have the following characteristics:

(a) The cutting edges should be free from wear or nicks.
(b) Both cutting edges should be the same angle and the same length.
(c) The margin should be free of wear.
(d) There should be a proper amount of lip clearance.

To Sharpen a Drill

A general-purpose drill has an included point angle of 118° and lip clearance from 8 to 12° (Fig. 14-14A, B).

1. Hold the drill near the point with one hand; with the other hand, hold the shank of the drill slightly lower than the point (Fig. 14-15).
2. Move the drill so that it is 59° to the face of the grinding wheel (Fig. 14-16).

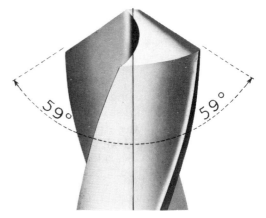

A - *Point angle*

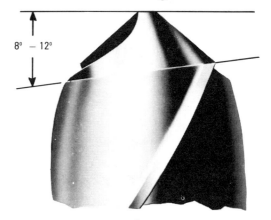

B - *Lip clearance*

Courtesy Cleveland Twist Drill (Canada) Ltd.

Fig. 14-14
The correct angle and clearance for a general purpose drill.

Courtesy Kostel Enterprises Ltd.

Fig. 14-15
The shank of the drill should be held lower than the point when sharpening.

A line scribed at 59° on the grinder work rest will help to keep the drill at the correct angle.

3. Have the lip or cutting edge of the drill parallel to the grinder tool rest.
4. Bring the lip of the drill against the grinding wheel and slowly lower the drill shank. DO NOT TWIST THE DRILL.

Courtesy Kostel Enterprises Ltd.

Fig. 14-16
The drill being held at 59° to the face of the grinding wheel.

5. Remove the drill from the wheel without moving the position of the body or hands, rotate the drill one-half turn, and grind the other cutting edge.
6. Check the angle of the drill point with a drill point gauge (Fig. 14-17).
7. Repeat operations 4 to 6 until the cutting edges are sharp and the lands are free from wear.

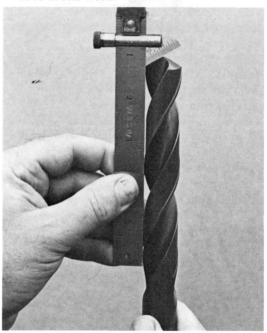

Courtesy Kostel Enterprises Ltd.

Fig. 14-17
Checking the drill point angle with a drill gauge.

ABRASIVE BELT GRINDER

Abrasive belt grinders have provided industry with a fast, easy, and economical method of finishing flat or contour work. In most cases, this machine has replaced the old hand method of using a file and abrasive cloth. Work that would take hours to finish by hand can be finished in a few minutes on an abrasive belt grinder.

The abrasive belt grinder, whether horizontal or vertical, consists of a motor,

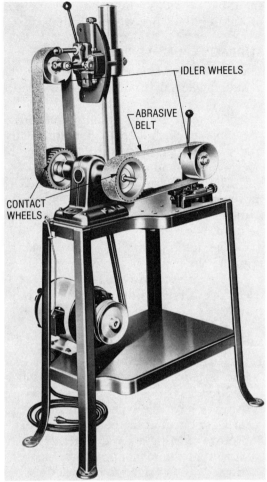

Courtesy Walker-Turner Division, Rockwell Manufacturing Co.

Fig. 14-18
The main parts of an abrasive belt grinder.

a contact wheel, an idler wheel, and an endless abrasive belt (Fig. 14-18).

ABRASIVE BELTS

Aluminum oxide and *silicon carbide* abrasive belts are used on belt grinders. Aluminum oxide abrasive belts should be used for grinding high-tensile strength materials (all steels and tough bronze). Silicon carbide abrasive belts should be used for grinding materials with low tensile strength (cast iron, aluminum, brass, copper, glass, plastic). A 60 to 80 grit belt may be used for general purpose work. For fine finishes, a 120 to 220 grit belt is recommended.

Safety Precautions

1. Always wear eye protectors when grinding.
2. Run the abrasive belts in the direction indicated by the arrows stamped on their backs.
3. Never grind on the up side of an abrasive belt (*i.e.*, with the belt rotating toward rather than away from you).
4. If much grinding or polishing is required on a workpiece, cool it frequently in a suitable medium.
5. Sharp corners or edges should be brought in contact with the belt lightly; otherwise these rough edges will tear the belt.

FINISHING FLAT SURFACES

Whenever flat surfaces are to be finished, a hard, flat platen (Fig. 14-19) should be mounted on the underside of the belt. This

Courtesy Norton Co. of Canada, Ltd.

Fig. 14-19
A platen under an abrasive belt allows flat surfaces to be finished on an abrasive belt grinder.

platen will prevent the belt from giving, thereby ensuring truly flat work.

FINISHING CONTOUR SURFACES

Concave and convex surfaces are easily finished by mounting a soft or formed contact wheel on the machine. The work is then held against the contact wheel, which conforms to the shape of the part (Fig. 14-20). This method makes it possible to finish intricate forms or sharp radii.

Courtesy Norton Co. of Canada, Ltd.

Fig. 14-20
A soft or formed contact wheel should be used for finishing contour surfaces.

SURFACE GRINDER

The surface grinder, used primarily for the grinding of flat surfaces, has become an important machine tool. It can be used to grind either hardened or unhardened workpieces to the close tolerances and high surface finishes required for many jobs.

The most common surface grinder is the horizontal spindle grinder with a reciprocating table (Fig. 14-21), which consists of a revolving abrasive wheel mounted on a horizontal spindle and a rectangular table which moves back and forth under the wheel. A magnetic chuck mounted on the table provides a fast and easy method of holding the work while grinding.

SURFACE GRINDER PARTS

The *wheel feed handwheel* moves the grinding wheel up or down to set the depth of cut.

The *table traverse handwheel* moves the table back and forth (longitudinally) under the grinding wheel. It can be operated by hand or automatically.

The *crossfeed handwheel* is used to move the table in or out (transversely). Either a hand or automatic crossfeed can be used.

The *table reverse dogs* are used to regulate the length of table travel.

The *table traverse reverse lever* is used to reverse the direction of the table travel.

Safety Precautions

1. Never run a grinding wheel faster than the speed recommended on the wheel blotter.
2. Always have the wheel guard covering at least one-half of the grinding wheel.
3. Before starting a grinder, *always* make sure that the magnetic chuck has been turned on, by trying to remove the work from the chuck (Fig. 14-22).

Courtesy Kostel Enterprises Ltd.

Fig. 14-22
Before starting a grinder, make sure that the magnetic chuck is on, by trying to remove the workpiece.

4. See that the wheel clears the work before starting the grinder.
5. Stand to one side of the wheel before starting the machine.
6. Never attempt to clean the magnetic chuck, or mount and remove work, until the wheel has completely stopped.
7. *ALWAYS* wear safety goggles while grinding.

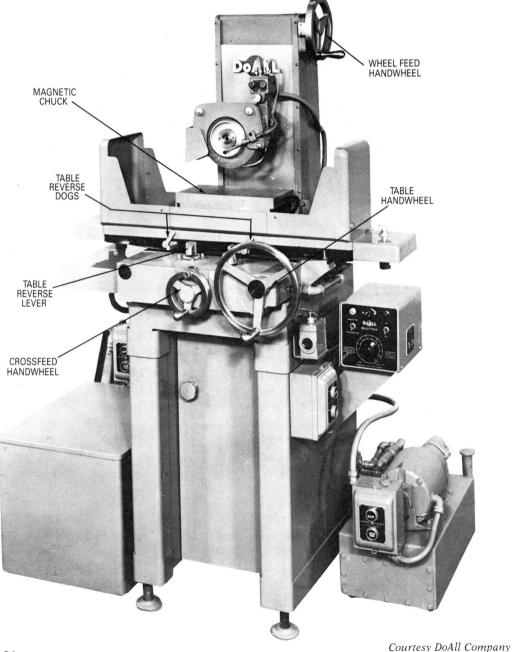

Fig. 14-21
The main parts of a surface grinder.

Courtesy DoAll Company

SURFACE GRINDER

Courtesy Kostel Enterprises Ltd.

Fig. 14-23
Test a grinding wheel before mounting, to make sure that it is not cracked or damaged.

MOUNTING A GRINDING WHEEL

A type No. 1 aluminum oxide straight wheel (Fig. 14-23) is generally used for most surface grinding operations. Before a wheel is mounted on a machine, it should be checked to make sure it is not defective. Suspend the wheel by slipping one finger through the hole and gently tap the side with the handle of a hammer or screwdriver (Fig. 14-23). A good wheel will give a sharp clear ring.

Care should be used when handling or mounting a grinding wheel, to prevent it from being damaged. A wheel that has been misused or damaged may shatter, cause damage to the grinder, and could cause a serious accident.

To Mount a Wheel on a Straight Spindle

1. Check that the wheel is not cracked, by tapping it at four points about 90° apart with a plastic or wooden handled screwdriver (Fig. 14-23).
   A good wheel will give a sharp, clear ring.
2. Clean the inner flange on the machine and the hole in the grinding wheel.
3. Check the wheel to make sure there is a wheel blotter on each side.
   If not, secure two blotters the same size as the flanges and place one on each side of the wheel.
4. Slide the wheel on the grinder spindle. The wheels should go on freely without binding. *Never force a wheel onto the spindle.*
5. Clean and place the outer flange against the wheel.
6. Hold the wheel with a rag and tighten the spindle nut against the flange enough to hold the wheel firmly (Fig. 14-24).
   DO NOT exert excessive pressure while tightening, or strains may be set up in the wheel that may cause it to break.
7. Replace the grinding wheel guard on the machine.

Truing and Dressing a Wheel

Truing is the operation of making a wheel run true, or altering the face to give it a desired shape.

Courtesy Kostell Enterprises Ltd.

Fig. 14-24
Tightening a grinding wheel on a straight spindle.

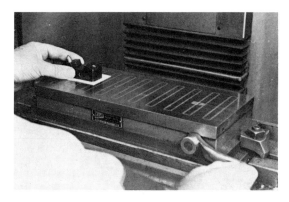

Courtesy Kostel Enterprises Ltd.

Fig. 14-25
Setting a diamond dresser on a magnetic chuck.

Dressing is the operation of removing dull grains or metal particles to make the wheel cut better. A dull or loaded wheel should be dressed for several reasons:
(a) to keep down the heat generated between the wheel and the work;
(b) to reduce the strain on the grinding wheel and machine;
(c) to improve the surface finish of the work.

On a surface grinder, an industrial diamond mounted in a holder is used to dress and true the wheel (Fig. 14-25).

To Dress and True a Surface Grinder Wheel

1. Check the diamond for wear, and when necessary, turn it in the holder to expose a new point.
2. The diamond is canted in the holder at a 10° angle.
   This helps to prevent chattering and the tendency to dig in during the dressing operation.
3. Clean the magnetic chuck thoroughly with a cloth, and then wipe over it with the palm of the hand.
4. Place the diamond on the last two magnets, on the left-hand end of the magnetic chuck.
   Paper should be placed between the diamond and the chuck to prevent scratching or marring the chuck surface when removing the diamond holder (Fig. 14-25).
5. The point of the diamond should be offset about 12 mm (1/2 in.) to the left of the grinding wheel centre line (Fig. 14-26).
6. Energize the magnetic chuck by turning the lever to the ON position (Fig. 14-25).
7. Make sure the diamond clears the wheel; then start the grinder.
8. Lower the wheel until it touches the diamond.
9. Move the diamond *slowly* across the face of the wheel.
10. Take light cuts 0.02 mm or 0.001 in. until the wheel is clean and sharp and is running true.
11. Take a finish pass of 0.01 mm or 0.0005 in. across the face of the grinding wheel.

Courtesy Kostel Enterprises Ltd.

Fig. 14-26
The diamond should be offset about 12 mm (1/2 in.) to the left of the wheel centre line for dressing.

GRINDING A FLAT SURFACE

The most common operation performed on a surface grinder is that of grinding a flat or horizontal surface. To obtain the best results, the correct type of wheel, properly dressed, should be used.

To Grind a Flat Surface

1. Thoroughly clean the magnetic chuck with a cloth, and then wipe it with the palm of the hand.
2. File off any burrs on the surface of the work that is to be placed on the magnetic chuck.
3. Place a piece of paper *slightly* larger than the workpiece in the centre of the chuck.
4. Place the work on the paper and turn on the magnetic chuck.
   Try to remove the work to make sure it is held securely (Fig. 14-27).
5. Set the table reverse dogs so that the centre of the grinding wheel clears each end of the work by approximately 25 mm (1 in.).
6. Set the crossfeed to advance approximately 0.80 mm to 1.0 mm or 0.030 to 0.040 in. at every table reversal.
7. Turn the crossfeed handwheel until the edge of the work overlaps the edge of the grinding wheel by about 3 mm or 1/8 in. (Fig. 14-28).
8. Turn the wheelfeed handwheel until the grinding wheel is about 0.80 mm or 1/32 in. above the work surface.

Courtesy Kostel Enterprises Ltd.

Fig. 14-28
Setting the grinding wheel to the surface of the work.

9. Start the grinder and lower the wheelhead until the wheel just sparks the work.
10. Raise the wheel about 0.12 mm or 0.005 in.
    The wheel may have been set on a low spot of the work.
11. Start the table travelling automatically, and feed the entire width of the work under the wheel to check for high spots.
12. Lower the wheel 0.05 to 0.07 mm or 0.002 to 0.003 in. for every cut until the surface is completed.
    NOTE: *Cutting fluid should be used whenever possible to aid the grinding action and keep the work cool.*

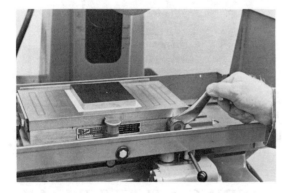

Courtesy Kostel Enterprises Ltd.

Fig. 14-27
Energizing the magnetic chuck to hold the workpiece.

TEST YOUR KNOWLEDGE
Bench or Pedestal Grinders

1. Define "offhand grinding".
2. Why is a coarse and a fine-grained wheel usually found on bench or pedestal grinders?

3. How close should the work rests be set to the wheel? Explain why.

Grinding Wheels
4. Name the two types of grinding wheels used on bench or pedestal grinders.
5. Name four natural abrasives and explain why they are rarely used in wheel manufacture.
6. What types of materials are ground with
   (a) aluminum oxide wheels?
   (b) silicon carbide wheels?

Grinder Safety
7. What precaution should be observed when starting a grinder?
8. Why should a new wheel be left running for about one minute before using?
9. Why should burrs produced by grinding be removed?

Dressing and Truing a Wheel
10. Name three reasons for dressing and truing a wheel.
11. Define "dressing" and "truing".
12. Name two types of wheel dressers.
13. Explain how to dress and true a grinding wheel.

Grinding a Lathe Toolbit
14. What are the end- and side-relief angles of a general-purpose lathe toolbit?
15. What instrument can be used to check the toolbit angles and clearances?
16. Why is it recommended to move the toolbit back and forth over the face of the grinding wheel?
17. Why should the cutting edge of a toolbit be honed after grinding?
18. How can a general-purpose toolbit be resharpened quickly and easily?

Drill Grinding
19. List the characteristics of a correctly ground drill.
20. What is the lip clearance and point angle of a general-purpose drill?
21. At what angle is the drill held to the face of the wheel?
22. How should the cutting edge be held in relation to the grinder tool rest?

Abrasive Belt Grinder
23. Why are abrasive belt grinders used in industry?

Abrasive Belts
24. Name two types of abrasive grinding belts.
25. For what purpose is each type used?

Safety Precautions
26. In what direction should abrasive belts be run?
27. How should sharp edges of work be brought in contact with the belt? Explain why.

Finishing Surfaces
28. Why is a platen used when finishing flat surfaces?
29. How can contour surfaces be finished?

Surface Grinder
30. Why is the surface grinder considered an important machine tool?
31. Briefly describe a common surface grinder.
32. What device is generally used to hold work while grinding?

Surface Grinder Parts
33. Describe the purpose of:
    (a) the crossfeed handwheel
    (b) the table traverse handwheel
    (c) the wheel feed handwheel

Safety Precautions
34. Before starting the grinder, how do you test the magnetic chuck for holding power?
35. Explain the importance of any three safety precautions for a surface grinder.

Mounting a Grinding Wheel
36. Explain the procedure for testing a wheel to make sure it is not defective.
37. List the steps to mount a grinding wheel.
38. Why should excessive pressure not be used when tightening the spindle nut?

Truing and Dressing a Wheel
39. Define: truing, dressing.
40. Why should a dull or loaded wheel be dressed?
41. Where should the diamond holder be placed on the magnetic chuck?
42. Why is paper used between the chuck and the diamond holder?
43. How should the diamond be located in relation to the wheel?

Grinding a Flat Surface
44. Explain the procedure for mounting work on a magnetic chuck.
45. How long should the table travel be in relation to the work length?
46. Explain the procedure for setting the wheel to the work surface.
47. Why should cutting fluid be used whenever possible?

CHAPTER 15
COMPUTER AGE MACHINING

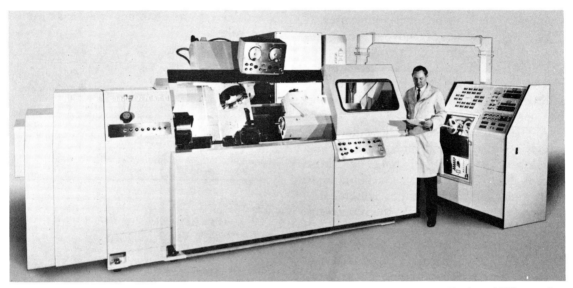

Courtesy Cincinnati Milacron Inc.

The development of the computer has made changes in our everyday life from retail sales, banking, and medicine to communications, transportation, science, and manufacturing. No other invention in history has had such an impact on humanity, in such a short period of time, as has the computer. Nothing in our life seems to be unaffected by the computer, which has made possible the exploration of space, world-wide television, improved health care, quality-controlled manufacturing, the use of robots, flexible manufacturing systems, and many others. Since our present-day computer is considered to be in its infancy, it is hard to imagine what effect the development of newer and more powerful computers will have. One thing is certain: this revolutionary invention will drastically change our lifestyles and the world in general.

HISTORY OF THE COMPUTER

Ever since the beginning of mankind, some type of device or system has been used to count and perform calculations. Over the ages there have been continual developments to improve the methods of counting and making calculations. Primitive people used their fingers, toes, and stones to count (Fig. 15-1). The *abacus* (Fig. 15-2),

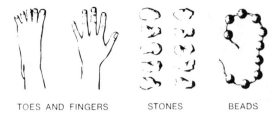

TOES AND FINGERS   STONES   BEADS

Fig. 15-1
The equipment used by primitive people for counting.

developed in the Orient around 4000 B.C., was really the first primitive computer. It uses the principle of moving beads on several wires to make calculations. The abacus is a very accurate and quick method of making calculations, and can still be found in use today in some Oriental businesses.

The first mechanical calculator, which could only add and subtract, was developed in France in 1642 by Blaise Pascal. In 1671 a German mathematician built a mechanical calculator which could also multiply and divide. In the nineteenth century Charles Babbage of England built a difference engine, which could rapidly and accurately calculate long lists of various functions, including logarithms. The punched card system, which was the first method of data processing, was introduced in 1804 by a Frenchman named J.M. Jacquard. The information was punched on cards (Fig. 15-3), and then read and tabulated by electronic sensors.

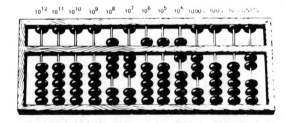

Fig. 15-2
The abacus, developed around 4000 B.C., was the first real computer.

The first simple computer, used to calculate wing designs for the aircraft industry, was developed in the 1930s by a German, Konrad Zuse. In 1939 George Stibitz produced a computer for the Bell Telephone Laboratories in the United States. This computer was capable of doing calculations over telephone wires, giving birth to the first remote data processing machine. During World War II the development of the computer progressed steadily and the early digital computers used electro-mechanical on-off switches and relays.

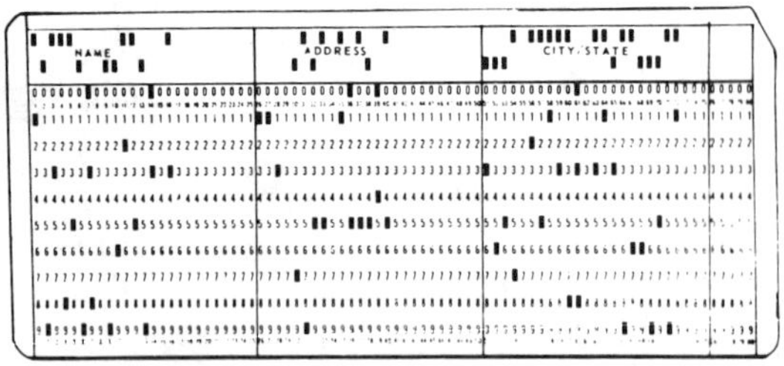

Fig. 15-3
Punched cards were our first method of data processing.

In 1946 the world's first electronic digital computer, the ENIAC (electronic numerical integrating automatic computer) was produced. It contained 19000 vacuum tubes, weighed 27 t (30 tons), and took up more than 1400 m² (15000 sq. ft.) of floor space. It was capable of adding two numbers in 1/5000 of a second. The transistor was developed in 1947 and quickly replaced the bulky vacuum tubes. The silicon chip (integrated circuits), developed in the late 1950s, allowed thousands of transistors and circuits to be incorporated in a tiny (about 6 mm [1/4 in.]) square piece of silicon (Fig. 15-4).

In 1971 the Intel Corporation developed the microprocessor, a chip which contained the entire *central processing unit* (CPU) for a simple computer. This

Courtesy Rockwell International

Fig. 15-4
Thousands of transistors and circuits are contained in a tiny silicon chip.

single chip could be programmed to do any number of tasks, from steering a spacecraft to operating a watch or controlling a personal computer. Developments in the future will continue to make new computers smaller and more powerful. The *information revolution*, which was predicted for many years, has now arrived. It will make dramatic changes in the way people live and work. Because of the computer revolution, productivity will increase and thereby raise the standard of living; the world will never be the same again.

THE ROLE OF THE COMPUTER

Computers will continue to affect our everyday lives. Some computers can perform one million calculations per second, making possible mathematical calculations which earlier would have taken months to do. Medical centres are cataloguing all known diseases, along with their symptoms and known cures. Children may learn more, at a younger age, because the computer can be used as a teaching tool. The computer can be used to predict weather forecasts and guide planes, missiles, and spacecraft. Retail stores use computers to record purchases, keep inventory up-to-date, record buying habits, and do all the necessary accounting and bookkeeping.

In the manufacturing world, computers have contributed greatly to the efficient manufacture of all goods. Computers are being used to operate and control machine tools at the maximum speeds possible, while at the same time assessing the accuracy of the product and taking corrective measures if necessary. *Computer-aided design* (CAD) can design a product on a screen and test and modify the design even before production has begun (Fig. 15-5). *Computer-aided manufacturing* (CAM) results in less scrap and provides more reliability through computer control of the machining sequences and cutting speeds and feeds.

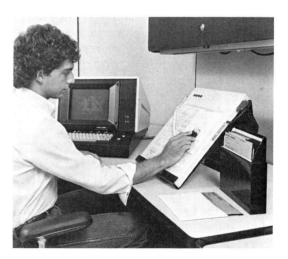

Courtesy Bausch & Lomb

Fig. 15-5
CAD systems are used by engineers who research and design products.

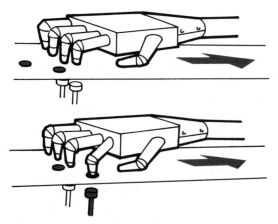

Fig. 15-6
Using mechanical wires to illustrate how information could be decoded from punched tape.

NUMERICAL CONTROL

Numerical control (NC) is one of the most exciting developments in the machine tool industry of the past century. Numerical control may be defined as a means of accurately controlling the movement of machine tools by a *series of programmed numerical data* which activates the motors of the machine tool. Numerical control is really an efficient way of reading prints and conveying this information to the machine's motors, which control the speeds, feeds, and various movements of the machine tool.

The designer or tool engineer's information on a print is coded and punched into a tape, which in turn is fed into the machine tool reader. The punched tape may be compared to the paper roll used to operate the keys of a player piano and produce music. Each hole on the paper roll corresponds to a key (note) on the player piano keyboard. As air passes through the hole, it moves a key on the keyboard and produces a musical note. On numerical control tape reading devices, beams of light are generally used to decode the information contained on the tape. Each time a hole in the tape appears under a beam of light, a specific circuit is activated, sending signals to start or stop motors (which position the work) and control the various functions of the machine tool. Figure 15-6 illustrates the principle of how information might be decoded from a numerical control tape by small mechanical wires (fingers).

The Electronics Industries Association has chosen the 25.4 mm (1 in.) 8-channel tape (Fig. 15-7) as the standard for the industry. It may be made of paper, foil, or mylar. Mylar (a type of plastic) is generally used, because it is very sturdy and almost impossible to tear.

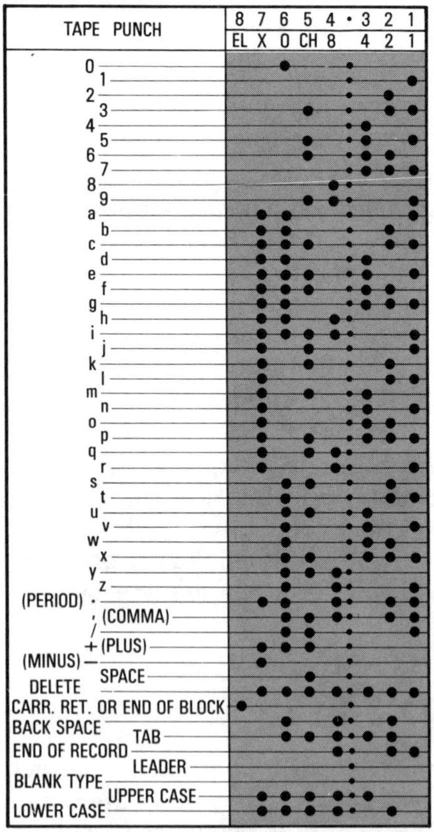

Courtesy Cincinnati Milacron Inc.

Fig. 15-7
Electronics Industries Association standard coding system for a 25.4 mm (1 in.) wide, 8-channel tape.

250 CHAPTER 15 / COMPUTER AGE MACHINING

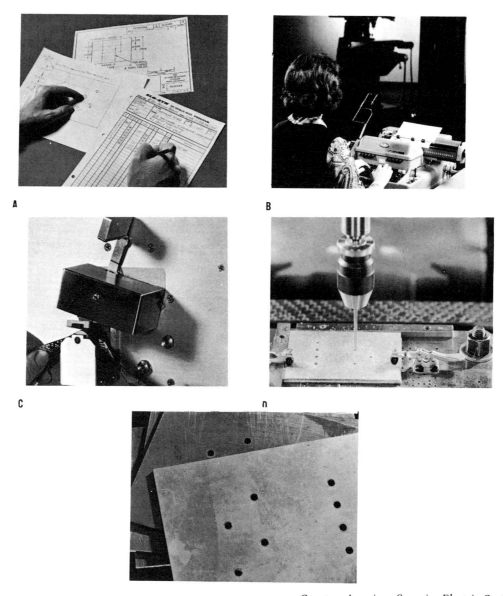

Courtesy American Superior Electric Co. Ltd.

Fig. 15-8
The numerical control sequence of operations, from programming to the finished part.

Numerical Control Operation

Although numerical control systems differ greatly in detail and complexity from manufacturer to manufacturer, all have basically the same elements. Regardless of the type of input media used, all operate a machine tool in the same way. A complete sequence of operations, beginning with the programmer transferring information from a print to a program sheet and continuing until the final part is produced on a machine, is illustrated in Fig. 15-8.

1. The programmer reviews the print of the part to be machined, determines the sequence of operations required, and

NUMERICAL CONTROL   251

lists the particulars about each operation on a program sheet (Fig. 15-8A).
2. A typist transfers the information contained on the program sheet to the tape, using a special tape punching machine (Fig. 15-8B). Generally two tapes are made by two different typists, and these are compared to ensure that no error has been made.
3. The punched tape and a copy of the program sheet are then handed to the machine operator. After positioning the part to be machined on the machine table according to the instructions contained on the program sheet, the tape is threaded into the tape reader (Fig. 15-8C).
4. The tape reader, which automatically advances the tape, is then started.
5. As each block of information is decoded by the tape reader, it sends the necessary information to the *cycle control*.
6. At the end of each machining operation, the *feedback switch* informs the cycle control that the previous operation has been completed (Fig. 15-8D).
7. The cycle control instructs the *command memory unit* to transfer the tape instructions for the next operation to the *indexer*.
8. The indexer starts the servo motors, which in turn move the machine table slides the required amount.

RECENT ADVANCES IN NUMERICAL CONTROL

With advancing technology, more efficient use of NC has been developed. Continuing developments in microelectronics and computer technology have made this possible by shrinking the size of machine control systems, while at the same time providing greater capacity and capabilities. Most machine tools manufactured today are controlled by computers or other computer-like devices such as programmable controllers.

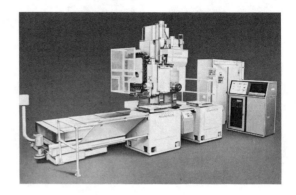

Courtesy Cincinnati Milacron Inc.

Fig. 15-9
A *computer-controlled machining centre.*

Computerized Numerical Control

Most of the NC systems being built today are computerized numerical control (CNC) systems. The CNC system, built around a powerful minicomputer (Fig. 15-9), incorporates a much larger memory capacity and has many more features to assist in programming, program editing at the machine, the setup of the machine, and its operation and maintenance. Many of these features are sets of machine and control instructions stored in memory, which can be called into use by the part program or by the machine operator.

Most CNC systems still have tape readers, with the part programs still being prepared in an office off-line unit and delivered to the machine in the form of punched tape. In some cases, however, the tape is read once, with part program being stored in memory for repetitive machining.

Now emerging are some CNC units that have more intelligence built into the computer. This permits the machine operator to input the data manually that describes the part to be machined. The computer then generates the part program, which is held in memory.

Direct Numerical Control

In this type of system, a number of CNC-equipped machines are controlled remotely from a mainframe computer. Most commonly, the direct numerical control (DNC) computer handles the scheduling of work and downloads into the machine's CNC memory a complete program when a new part is to be machined. It bypasses the control's tape reader. If automatic work-handling is provided at the machines, the computer can control the automatic cycling of each machine. In this case, no machine operator is needed except for initial setup or trouble shooting.

Computer-Aided Design

The computer has also found great use in the process of designing parts and components. Computer-aided design (CAD), a new system introduced in the 1960s, allows the designer or engineer to produce finished parts or engineering drawings from simple pencil sketches or models (Fig. 15-10). It also allows the designer an opportunity to quickly make design changes in the part if it appears not to be functional. From three-drawing orthographic views, it is possible to transform these drawings into a three-dimensional view and, with appropriate computer software, to show how a part would perform in use. This has given designers a real advantage; they can now test and change the design until the part works properly, instead of going through the costly process of actually making a part first.

If a product such as an airplane is being designed, each part can be designed using the CAD system, and changes can be made wherever necessary. All these parts can then be assembled on the video screen, and the entire unit can be tested for performance. If changes have to be made, it is possible to make them in a few minutes with the CAD system, where previously these changes would have taken months to complete using the standard procedures.

TURNING AND CHUCKING CENTRES

About 40 percent of all metal-cutting operations consist of machining round work. Until the mid-1960s this work was produced on engine and turret lathes, which were not very efficient by present-day standards. Intensive research led to the development of numerically controlled turning centres for machining work between centres, and chucking centres for machining work held in a chuck. These turning and chucking centres, which are

Courtesy Cincinnati Milacron Inc.

Fig. 15-10
CAD systems are invaluable to engineers for designing products.

Courtesy Cincinnati Milacron Inc.

Fig. 15-11
A computer-controlled turning centre.

Courtesy Cincinnati Milacron Inc.

Fig. 15-12
Both turrets being used to machine a shaft having different diameters.

computer controlled, can machine almost any size and shape of work to very accurate dimensions and at very high production rates.

Turning Centres

Turning centres (Fig. 15-11) are designed mainly for machining shaft-type workpieces, which are held in a chuck and supported by a heavy-duty tailstock centre. On four-axis machines, two turrets, each holding seven different cutting tools, are mounted on separate cross-slides, one above and one below the centreline of the work. Because the turrets balance each other's cutting force, extremely heavy cuts can be taken from a workpiece supported by the tailstock centre (Fig. 15-12). The advantages of the two turret machines are:
(a) roughing and finishing cuts can be taken in one pass;
(b) different diameters can be machined at the same time;
(c) finishing and threading operations can be done at the same time.

The lower turret can also hold cutting tools for machining internal diameters on work held in a chuck.

Chucking Centres

Computer-controlled chucking centres are designed for machining most work held in a chuck or similar holding devices. The four-axis chucking centre has two turrets mounted on separate cross-slides, each holding up to seven cutting tools. While the upper turret is machining the inside diameter, the lower turret can machine the outside diameter (Fig. 15-13). However, both turrets can be used for internal cutting operations at the same time if the workpiece allows, or both can be used for machining outside diameters.

Courtesy Cincinnati Milacron Inc.

Fig. 15-13
The inside and outside diameters of a part being machined at the same time.

CHAPTER 16
HEAT TREATING

Courtesy Sunbeam Equipment Corporation

HEAT TREATMENT

Heat treatment is a term applied to a variety of procedures used to change the physical characteristics of metal by heating and cooling. Heat treatment is used to improve the microstructure of steel to meet certain physical specifications. Toughness, hardness, and wear resistance are a few of the qualities obtained through heat treatment. To obtain these characteristics, operations such as hardening, tempering, annealing, and case-hardening are necessary.

TYPES OF FURNACES

There are three types of furnaces used in various heat treating operations. These are the *low-temperature* or *drawing* furnace (Fig. 16-1A), the *high-temperature* furnace (Fig. 16-1B), and the *pot type* furnace (Fig.

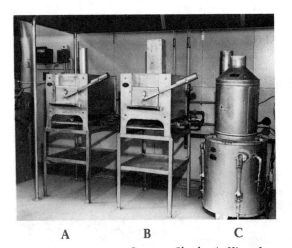

A   B   C

Courtesy Charles A. Hines Inc.

Fig. 16-1
Types of heat treating furnaces.

the pryometer, generally mounted on a wall near the furnace. Any changes in the furnace temperature show on the pyrometer, which may be set to shut off the heat to the furnace when a preset temperature is reached.

Another method used to determine the temperature of steel is to note its colour. This method is not as accurate as the pyrometer, but it can be used when heat treating small parts and tools. In Table 16-1 various heat colours and their approximate temperatures are given.

16-1C). The type of heat treating operation will determine the type of furnace used. For hardening carbon and alloy steels, and for pack carburizing machine steel, the high-temperature furnace is used. For preheating and tempering operations, the low-temperature furnace is used. Casehardening or liquid carburizing is carried out in the pot furnace. These furnaces may be heated by gas, oil, or electricity. The furnace temperature in all these furnaces is indicated and accurately controlled by a *thermocouple* and *pyrometer* (Fig. 16-2).

The *thermocouple*, which is made of two dissimilar metals, is mounted inside the furnace and is connected by leads to

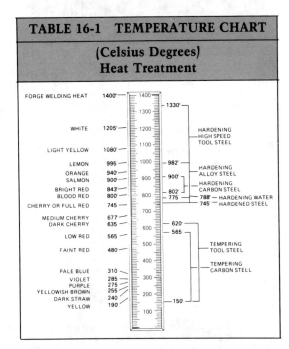

TABLE 16-1 TEMPERATURE CHART
(Celsius Degrees)
Heat Treatment

Colour	°C		Heat Treatment
FORGE WELDING HEAT	1400°	1400	
		1330°	
		1300	
WHITE	1205°	1200	HARDENING HIGH SPEED TOOL STEEL
		1100	
LIGHT YELLOW	1080°		
LEMON	995°	1000	982°
ORANGE	940°		HARDENING ALLOY STEEL
SALMON	900°	900	900°
BRIGHT RED	843°		HARDENING CARBON STEEL
BLOOD RED	800°	800	802°
			788° — HARDENING WATER
CHERRY OR FULL RED	745°		745° — HARDENED STEEL
MEDIUM CHERRY	677°	700	
DARK CHERRY	635°	600	620°
LOW RED	565°		565°
		500	TEMPERING TOOL STEEL
FAINT RED	480°	400	
			TEMPERING CARBON STEEL
PALE BLUE	310°	300	
VIOLET	285°		
PURPLE	275°		
YELLOWISH BROWN	255°	200	
DARK STRAW	240°		150°
YELLOW	190°	100	

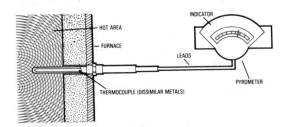

Fig. 16-2
The pyrometer and thermocouple accurately control and indicate furnace temperature.

SAFETY PRECAUTIONS

During any heat treating operation, the following safety precautions should be observed.

1. Know and understand the operation of the heat treating system.
2. Check that all the safety devices, such as automatic shut-off valves, air switches, exhaust fan, etc., are working properly before lighting the furnace.

3. Follow the manufacturer's instructions on lighting the furnace.
4. Stand to one side when lighting a gas- or oil-fired furnace, in case of a blow back.
5. Always wear a face shield, gloves, and approved protective clothing when working with hot metal.
6. Use the proper tongs for the job, and be sure the tongs are dry before removing any work from a liquid carburizing pot.

CAUTION: Any moisture (water or oil) which comes in contact with this liquid may cause an explosion and seriously burn the operator.

7. Never inhale the fumes from a liquid carburizing solution.

CAUTION: These solutions may contain CYANIDE OF POTASSIUM, WHICH IS A DEADLY POISON.

HARDENING CARBON STEEL

Hardening is the process of heating metal uniformly to its proper temperature and then quenching or cooling it in water, oil, air, or in a refrigerated area.

Carbon tool steels may be hardened by heating them to a bright cherry red colour, approximately 790 to 830°C (1450 to 1500°F), and quenching in water or oil. When the steel is heated to this temperature, a chemical and physical change takes place, with some of the carbon combining with the iron to form a new structure called *austenite*. When the steel is quenched, the austenite changes to a hard, brittle, fine-grained structure called *martensite* (Fig. 16-3).

To Harden Carbon Tool Steel

1. Light the furnace and preheat the steel slowly.
2. Heat the steel to its recommended temperature, checking either by colour or by pyrometer.

A - *Before hardening*   B - *After hardening*

Courtesy Kostel Enterprises Ltd.

Fig. 16-3
The grain structure of carbon tool steel.

3. Quench in water, brine, or oil, depending on the type of steel or manufacturer's recommendation.

NOTE: Long, slender pieces should be quenched vertically to avoid warping.

4. Move the work about in the quenching medium in a "figure 8" motion, to allow the steel to cool quickly and evenly.
5. Test for hardness with the edge of a file or a hardness tester.

TEMPERING

After hardening, the steel is brittle and may break with the slightest tap, due to the stresses caused by quenching. To overcome this brittleness, the steel is tempered; that is, it is reheated until it is brought to the desired temperature or colour, and then quenched. Tempering toughens the steel and makes it less brittle, although a little of the hardness is lost. As steel is heated it changes in colour, and these colours indicate various tempering temperatures (Table 16-2).

To Temper Carbon Tool Steel

1. Clean off all the surface scale from the work with abrasive cloth.
2. Select the heat or colour desired (Table 16-2).
3. Heat the steel slowly and evenly.
4. When the steel reaches the desired col-

| TABLE 16-2 TEMPERING COLOURS ||||
Colour	°C	°F	Tools
Faint straw	220	430	Toolbits, drills, taps
Medium straw	240	460	Punches and dies, milling cutters
Dark straw	255	490	Shear blades, hammer faces
Purple	270	520	Axes, wood chisels, tools
Dark blue	300	570	Knives, steel chisels
Light blue	320	610	Screwdrivers, springs

our or temperature, quench it quickly in the same cooling medium used for hardening.

ANNEALING

Annealing is the process of relieving internal strains and softening steel by heating it above its critical temperature (see Table 16-1), and allowing it to cool slowly in a closed furnace, or in ashes, lime, or asbestos.

CASEHARDENING

Casehardening is a method used to harden the outer surface of low-carbon steel while leaving the centre or core soft and ductile (Fig. 16-4). As carbon is the hardening agent, some method must be used to increase the carbon content of low-carbon steel before it can be hardened. Casehardening involves heating the metal to its critical temperature in some carbonaceous material. Three common methods used for casehardening are the pack method, the liquid bath method, and the gas method.

Pack method, also called *carburizing*, consists of packing the steel in a closed box with a carbonaceous material and heating it to a temperature of 900 to 927°C (1650 to 1700°F) for a period of 4 to 6 hours. The steel may then be removed from the box and quickly quenched in water or brine. To avoid excessive warping, it is sometimes better to allow the box to cool, remove the steel, reheat to 760 to 815°C (1400 to 1500°F), and then quench.

In school shops *Kasenit*, a non-poisonous coke compound, is often used for casehardening (pack method).
1. Heat the piece of steel to about 790 to 815°C (1450 to 1500°F), which will produce a bright cherry red colour.
2. Remove from furnace, and cover the work with powdered Kasenit. Replace metal in furnace and leave it there until the Kasenit appears to boil; then quench it in cold water.

This will only give from 0.25 to 0.40 mm (0.010 to 0.015 in.) penetration.

In the *liquid bath method*, the steel to be casehardened is immersed in a liquid cyanide bath containing up to 25% sodium cyanide. Potassium cyanide may also be

A - *Before casehardening*   B - *After casehardening*

Courtesy Kostel Enterprises Ltd.

Fig. 16-4
The grain structure of low-carbon steel.

used, but its fumes are dangerous. The temperature is held at 845°C (1550°F) for 15 minutes to 1 hour, depending on the depth of case required. At this temperature the steel will absorb both carbon and nitrogen from the cyanide. The steel is then quickly quenched in water or brine.

CAUTION: Never inhale the poisonous cyanide fumes or allow any water to come in contact with the cyanide, or explosive spattering may occur.

With the *gas method*, carburizing gases are used to caseharden low-carbon steel. The steel is placed in a furnace and sealed. Carburizing gas is introduced and the furnace is held between 900 and 927°C (1650 to 1700°F). After a predetermined time, the carburizing gas is shut off and the steel allowed to cool. The steel is then reheated to between 760 and 815°C (1400 to 1500°F) and quickly quenched in water or brine.

TESTING STEEL FOR HARDNESS

Hardness in steel may be defined as the property it has to resist penetration and deformation. The harder the steel, the greater its resistance to penetration and deformation. A common method of testing the hardness of steel is by trial and error, using a file, often called the *File Test*. Modern science has introduced many up-to-date methods that will test the hardness of metals very accurately. Some of these hardness testers are:

ROCKWELL The Rockwell hardness tester (Fig. 16-5) measures the amount of penetration caused by a diamond point being forced into the metal. The greater the penetration, the softer the metal. The penetration is recorded on a visible dial, and the resultant figure is called the Rockwell hardness number.

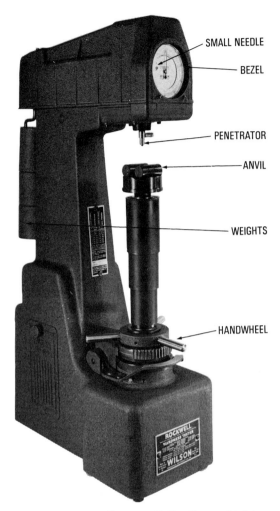

Courtesy Walker-Turner Division, Rockwell Manufacturing Co.

Fig. 16-5
A Rockwell hardness test can measure the hardness of a metal.

BRINELL This tester uses a round ball that is forced into the work surface. The diameter of the impression is measured by a special microscope. The higher the Brinell number, the harder the metal.

SHORE SCLEROSCOPE The test is done with a diamond-tipped hammer that is dropped on the metal from a given height. The amount of rebound indicates the degree of hardness.

HEAT TREATMENT SPECIFICATIONS FOR STEELS

There are two standard steel identifications: S.A.E. (Society of Automotive Engineers) and A.I.S.I. (American Iron and Steel Institute). To obtain the heat treatment specifications for various steels, use a standard reference book and look at the A.I.S.I. or S.A.E. tables of steels, or refer to the steel manufacturer's specifications.

FORGING AND SHAPING METAL

Forging is a process of shaping and forming steel by hammering. Bending, drawing, and shaping are some of the operations that can be done on steel when it is heated to its forging temperature. Most steels can be easily forged when they are a bright cherry red colour (approximately 790 to 815°C [450 to 1500°F]).

To Bend Steel

1. Lay out and centre-punch where the bend is to be made.
2. Light the furnace, and adjust the combustion if it is a gas furnace.

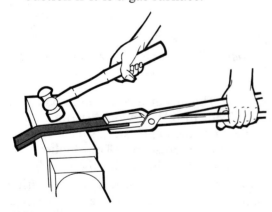

Fig. 16-6
Metal will bend readily when it is heated to a cherry red colour.

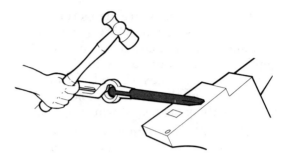

Fig. 16-7
Steel should be kept at a red heat when forging.

3. Place the work in the furnace and heat it to the proper temperature (approximately 790 to 815°C [1450 to 1500°F]). If an acetylene torch is available, it may be used to heat only the section where the bend is to be made.
4. Grip the work with a suitable pair of tongs and place it on the anvil. The centre-punch mark should be placed even with the edge of the anvil.
5. Bend the work over the face of the anvil by striking sharp blows on the end projecting over the anvil (Fig. 16-6).

To Forge Steel

1. Place the work in the furnace so that the complete part to be forged will heat evenly.
2. Allow the work to soak until it is bright red and the shadows have disappeared.
3. With the proper tongs, grip the work and hammer to the desired shape (Fig. 16-7).
   Do not forge the work when it cools to a dull red. Reheat to its forging temperature before finishing the shaping.
4. After forging, reheat the metal to its forging temperature and cover it in a box with lime, asbestos, or dry ashes, and allow it to cool slowly. The steel will then be annealed (softened).

TEST YOUR KNOWLEDGE

1. Define the term "heat treating."
2. Name three qualities obtained through heat treating.
3. List three heat treating operations.

Types of Furnaces

4. Name the three common types of heat treating furnaces, and state where each is used.
5. What fuel may be used to fire the various furnaces?
6. For what purpose is a pyrometer used?
7. Describe a thermocouple and explain its use.
8. If a pyrometer is not available, what other method can be used to check the temperature of the workpiece?
9. What temperatures do the following colours represent? Light yellow, bright cherry red, dark straw.

Safety Precautions

10. State five precautions to observe when heat treating steel.

Hardening and Tempering Carbon Steel

11. Define "hardening".
12. What change occurs to steel when it is hardened?
13. Define "martensite".
14. In point form, describe the procedure for hardening carbon tool steel.
15. How should long slender pieces be quenched? Explain why.
16. Why is it necessary to temper steel after hardening?
17. At what temperature should steel chisels be tempered?
18. List the procedure for tempering carbon tool steel.

Annealing

19. Define annealing.
20. Explain how a piece of hardened tool steel can be annealed.

Casehardening

21. What type of steel must be casehardened? Explain why.
22. Describe a piece of casehardened steel.
23. Name three methods of casehardening.
24. List in point form how a piece of low-carbon steel may be casehardened to a depth of 0.25 mm (0.010 in.).
25. Briefly explain how to carburize steel.

Testing Steel for Hardness

26. Name three methods of testing steel for hardness.
27. How does a Rockwell tester operate?

Forging and Shaping Metal

28. Define "forging".
29. At what temperature should the forging operation be performed?
30. When should forging be discontinued?

APPENDIX TABLES

TABLE 1 DECIMAL INCH, FRACTIONAL INCH, AND MILLIMETRE EQUIVALENTS

Decimal inch	Fractional inch	Millimetre	Decimal inch	Fractional inch	Millimetre
0.015625	1/64	0.397	0.515625	33/64	13.097
0.03125	1/32	0.794	0.53125	17/32	13.494
0.046875	3/64	1.191	0.546875	35/64	13.891
0.0625	1/16	1.588	0.5625	9/16	14.288
0.078125	5/64	1.984	0.578125	37/64	14.684
0.09375	3/32	2.381	0.59375	19/32	15.081
0.109375	7/64	2.778	0.609375	39/64	15.478
0.125	1/8	3.175	0.625	5/8	15.875
0.140625	9/64	3.572	0.640625	41/64	16.272
0.15625	5/32	3.969	0.65625	21/32	16.669
0.171875	11/64	4.366	0.671875	43/64	17.066
0.1875	3/16	4.762	0.6875	11/16	17.462
0.203125	13/64	5.159	0.703125	45/64	17.859
0.21875	7/32	5.556	0.71875	23/32	18.256
0.234375	15/64	5.953	0.734375	47/64	18.653
0.25	1/4	6.350	0.75	3/4	19.05
0.265625	17/64	6.747	0.765625	49/64	19.447
0.28125	9/32	7.144	0.78125	25/32	19.844
0.296875	19/64	7.541	0.796875	51/64	20.241
0.3125	5/16	7.938	0.8125	13/16	20.638
0.328125	21/64	8.334	0.828125	53/64	21.034
0.34375	11/32	8.731	0.84375	27/32	21.431
0.359375	23/64	9.128	0.859375	55/64	21.828
0.375	3/8	9.525	0.875	7/8	22.225
0.390625	25/64	9.922	0.890625	57/64	22.622
0.40625	13/32	10.319	0.90625	29/32	23.019
0.421875	27/64	10.716	0.921875	59/64	23.416
0.4375	7/16	11.112	0.9375	15/16	23.812
0.453125	29/64	11.509	0.953125	61/64	24.209
0.46875	15/32	11.906	0.96875	31/32	24.606
0.484375	31/64	12.303	0.984375	63/64	25.003
0.5	1/2	12.700	1.	1	25.400

TABLE 2

Conversion of Inches to Millimetres						Conversion of Millimetres to Inches					
Inches	Millimetres	Inches	Millimetres	Inches	Millimetres	Millimetres	Inches	Millimetres	Inches	Millimetres	Inches
0.001	0.025	0.290	7.37	0.660	16.76	0.01	0.0004	0.35	0.0138	0.68	0.0268
0.002	0.051	0.300	7.62	0.670	17.02	0.02	0.0008	0.36	0.0142	0.69	0.0272
0.003	0.076	0.310	7.87	0.680	17.27	0.03	0.0012	0.37	0.0146	0.70	0.0276
0.004	0.102	0.320	8.13	0.690	17.53	0.04	0.0016	0.38	0.0150	0.71	0.0280
0.005	0.127	0.330	8.38	0.700	17.78	0.05	0.0020	0.39	0.0154	0.72	0.0283
0.006	0.152	0.340	8.64	0.710	18.03	0.06	0.0024	0.40	0.0157	0.73	0.0287
0.007	0.178	0.350	8.89	0.720	18.29	0.07	0.0028	0.41	0.0161	0.74	0.0291
0.008	0.203	0.360	9.14	0.730	18.54	0.08	0.0031	0.42	0.0165	0.75	0.0295
0.009	0.229	0.370	9.40	0.740	18.80	0.09	0.0035	0.43	0.0169	0.76	0.0299
0.010	0.254	0.380	9.65	0.750	19.05	0.10	0.0039	0.44	0.0173	0.77	0.0303
0.020	0.508	0.390	9.91	0.760	19.30	0.11	0.0043	0.45	0.0177	0.78	0.0307
0.030	0.762	0.400	10.16	0.770	19.56	0.12	0.0047	0.46	0.0181	0.79	0.0311
0.040	1.016	0.410	10.41	0.780	19.81	0.13	0.0051	0.47	0.0185	0.80	0.0315
0.050	1.270	0.420	10.67	0.790	20.07	0.14	0.0055	0.48	0.0189	0.81	0.0319
0.060	1.524	0.430	10.92	0.800	20.32	0.15	0.0059	0.49	0.0193	0.82	0.0323
0.070	1.778	0.440	11.18	0.810	20.57	0.16	0.0063	0.50	0.0197	0.83	0.0327
0.080	2.032	0.450	11.43	0.820	20.83	0.17	0.0067	0.51	0.0201	0.84	0.0331
0.090	2.286	0.460	11.68	0.830	21.08	0.18	0.0071	0.52	0.0205	0.85	0.0335
0.100	2.540	0.470	11.94	0.840	21.34	0.19	0.0075	0.53	0.0209	0.86	0.0339
0.110	2.794	0.480	12.19	0.850	21.59	0.20	0.0079	0.54	0.0213	0.87	0.0343
0.120	3.048	0.490	12.45	0.860	21.84	0.21	0.0083	0.55	0.0217	0.88	0.0346
0.130	3.302	0.500	12.70	0.870	22.10	0.22	0.0087	0.56	0.0220	0.89	0.0350
0.140	3.56	0.510	12.95	0.880	22.35	0.23	0.0091	0.57	0.0224	0.90	0.0354
0.150	3.81	0.520	13.21	0.890	22.61	0.24	0.0094	0.58	0.0228	0.91	0.0358
0.160	4.06	0.530	13.46	0.900	22.86	0.25	0.0098	0.59	0.0232	0.92	0.0362
0.170	4.32	0.540	13.72	0.910	23.11	0.26	0.0102	0.60	0.0236	0.93	0.0366
0.180	4.57	0.550	13.97	0.920	23.37	0.27	0.0106	0.61	0.0240	0.94	0.0370
0.190	4.83	0.560	14.22	0.930	23.62	0.28	0.0110	0.62	0.0244	0.95	0.0374
0.200	5.08	0.570	14.48	0.940	23.88	0.29	0.0114	0.63	0.0248	0.96	0.0378
0.210	5.33	0.580	14.73	0.950	24.13	0.30	0.0118	0.64	0.0252	0.97	0.0382
0.220	5.59	0.590	14.99	0.960	24.38	0.31	0.0122	0.65	0.0256	0.98	0.0386
0.230	5.84	0.600	15.24	0.970	24.64	0.32	0.0126	0.66	0.0260	0.99	0.0390
0.240	6.10	0.610	15.49	0.980	24.89	0.33	0.0130	0.67	0.0264	1.00	0.0394
0.250	6.35	0.620	15.75	0.990	25.15	0.34	0.0134	....	....	....	....
0.260	6.60	0.630	16.00	1.000	25.40						
0.270	6.86	0.640	16.26	....	....						
0.280	7.11	0.650	16.51	....	....						

Courtesy Automatic Electric Company

TABLE 3 LETTER DRILL SIZES

Letter	mm	in.	Letter	mm	in.	Letter	mm	in.	Letter	mm	in.
A	5.9	0.234	H	6.7	0.266	N	7.7	0.302	T	9.1	0.358
B	6.0	0.238	I	6.9	0.272	O	8.0	0.316	U	9.3	0.368
C	6.1	0.242	J	7.0	0.277	P	8.2	0.323	V	9.5	0.377
D	6.2	0.246	K	7.1	0.281	Q	8.4	0.332	W	9.8	0.386
E	6.4	0.250	L	7.4	0.290	R	8.6	0.339	X	10.1	0.397
F	6.5	0.257	M	7.5	0.295	S	8.8	0.348	Y	10.3	0.404
G	6.6	0.261							Z	10.5	0.413

TABLE 4 NUMBER DRILL SIZES

No.	mm	inch	No.	mm	inch	No.	mm	inch
1	5.80	0.2280	34	2.81	0.1110	66	0.84	0.0330
2	5.60	0.2210	35	2.79	0.1100	67	0.81	0.0320
3	5.40	0.2130	36	2.70	0.1065	68	0.79	0.0310
4	5.30	0.2090	37	2.65	0.1040	69	0.74	0.0292
5	5.22	0.2055	38	2.60	0.1015	70	0.71	0.0280
6	5.18	0.2040	39	2.55	0.0995	71	0.66	0.0260
7	5.10	0.2010	40	2.50	0.0980	72	0.64	0.0250
8	5.05	0.1990	41	2.45	0.0960	73	0.61	0.0240
9	5.00	0.1960	42	2.40	0.0935	74	0.57	0.0225
10	4.91	0.1935	43	2.25	0.0890	75	0.53	0.0210
11	4.85	0.1910	44	2.20	0.0860	76	0.51	0.0200
12	4.80	0.1890	45	2.10	0.0820	77	0.46	0.0180
13	4.70	0.1850	46	2.05	0.0810	78	0.41	0.0160
14	4.62	0.1820	47	2.00	0.0785	79	0.37	0.0145
15	4.57	0.1800	48	1.95	0.0760	80	0.34	0.0135
16	4.50	0.1770	49	1.85	0.0730	81	0.33	0.0130
17	4.40	0.1730	50	1.80	0.0700	82	0.32	0.0125
18	4.30	0.1695	51	1.70	0.0670	83	0.31	0.0120
19	4.20	0.1660	52	1.60	0.0635	84	0.29	0.0115
20	4.10	0.1610	53	1.50	0.0595	85	0.28	0.0110
21	4.03	0.1590	54	1.40	0.0550	86	0.27	0.0105
22	4.00	0.1570	55	1.30	0.0520	87	0.25	0.0100
23	3.91	0.1540	56	1.20	0.0465	88	0.24	0.0095
24	3.86	0.1520	57	1.10	0.0430	89	0.23	0.0091
25	3.80	0.1495	58	1.06	0.0420	90	0.22	0.0087
26	3.73	0.1470	59	1.04	0.0410	91	0.21	0.0083
27	3.65	0.1440	60	1.00	0.0400	92	0.20	0.0079
28	3.60	0.1405	61	0.99	0.0390	93	0.19	0.0075
29	3.50	0.1360	62	0.97	0.0380	94	0.18	0.0071
30	3.30	0.1285	63	0.94	0.0370	95	0.17	0.0067
31	3.00	0.1200	64	0.92	0.0360	96	0.16	0.0063
32	2.95	0.1160	65	0.89	0.0350	97	0.15	0.0059
33	2.85	0.1130						

TABLE 5 COMMERCIAL TAP DRILL SIZES (75% of thread depth) AMERICAN NATIONAL AND UNIFIED FORM THREAD

\multicolumn{3}{c}{NC National Coarse}			NF National Fine		
Tap Size	Threads per inch	Tap Drill Size	Tap Size	Threads per inch	Tap Drill Size
# 5	40	#38	# 5	44	#37
# 6	32	#36	# 6	40	#33
# 8	32	#29	# 8	36	#29
#10	24	#25	#10	32	#21
#12	24	#16	#12	28	#14
1/4	20	# 7	1/4	28	# 3
5/16	18	F	5/16	24	I
3/8	16	5/16	3/8	24	Q
7/16	14	U	7/16	20	25/64
1/2	13	27/64	1/2	20	29/64
9/16	12	31/64	9/16	18	33/64
5/8	11	17/32	5/8	18	37/64
3/4	10	21/32	3/4	16	11/16
7/8	9	49/64	7/8	14	13/16
1	8	7/8	1	14	15/16
1-1/8	7	63/64	1-1/8	12	1-3/64
1-1/4	7	1-7/64	1-1/4	12	1-11/64
1-3/8	6	1-7/32	1-3/8	12	1-19/64
1-1/2	6	1-11/32	1-1/2	12	1-27/64
1-3/4	5	1-9/16			
2	4-1/2	1-25/32			

NPT NATIONAL PIPE THREAD

Tap Size	Threads per inch	Tap Drill Size	Tap Size	Threads per inch	Tap Drill Size
1/8	27	11/32	1	11-1/2	1-5/32
1/4	18	7/16	1-1/4	11-1/2	1-1/2
3/8	18	19/32	1-1/2	11-1/2	1-23/32
1/2	14	23/32	2	11-1/2	2-3/16
3/4	14	15/16	2-1/2	8	2-5/8

The major diameter of an NC or NF number size tap or screw = (N × 0.013) + 0.060
EXAMPLE: The major diameter of a #5 tap equals
   (5 × 0.013) + 0.060 = 0.125 diameter

TABLE 6 METRIC TAP DRILL SIZES

Nominal Diameter mm	Thread Pitch mm	Tap Drill Size mm	Nominal Diameter mm	Thread Pitch mm	Tap Drill Size mm
1.60	0.35	1.20	20.00	2.50	17.50
2.00	0.40	1.60	24.00	3.00	21.00
2.50	0.45	2.05	30.00	3.50	26.50
3.00	0.50	2.50	36.00	4.00	32.00
3.50	0.60	2.90	42.00	4.50	37.50
4.00	0.70	3.30	48.00	5.00	43.00
5.00	0.80	4.20	56.00	5.50	50.50
6.00	1.00	5.30	64.00	6.00	58.00
8.00	1.25	6.80	72.00	6.00	66.00
10.00	1.50	8.50	80.00	6.00	74.00
12.00	1.75	10.20	90.00	6.00	84.00
14.00	2.00	12.00	100.00	6.00	94.00
16.00	2.00	14.00			

TABLE 7
ISO METRIC PITCH & DIAMETER COMBINATIONS

Nominal Dia. (mm)	Thread Pitch (mm)	Nominal Dia. (mm)	Thread Pitch (mm)
1.6	0.35	20	2.5
2	0.40	24	3.0
2.5	0.45	30	3.5
3	0.50	36	4.0
3.5	0.60	42	4.5
4	0.70	48	5.0
5	0.80	56	5.5
6	1.00	64	6.0
8	1.25	72	6.0
10	1.50	80	6.0
12	1.75	90	6.0
14	2.00	100	6.0
16	2.00		

GLOSSARY

ABRASIVE — The material used in making grinding wheels or abrasive cloth; it may be either natural or artificial. The natural abrasives are emery and corundum. The artificial abrasives are silicon carbide and aluminum oxide.

ALIGNMENT — Linear accuracy, uniformity, or coincidence of the centres of a lathe; a straight line of adjustment through two or more points. Setting the lathe in alignment means adjusting the tailstock in line with the headstock spindle to produce parallel work.

ALLOWANCE — As applied to the fitting of machine parts, allowance means a difference in dimensions prescribed in order to secure classes of fits. It is the minimum or maximum interference intentionally permitted between mating parts.

ALLOY — A mixture of two or more metals melted together. As a rule, when two or more metals are melted together to form an alloy, the substance formed is a new metal.

ALUMINUM — A very light silvery-white metal used independently or in alloys with copper and other metals.

ARBOR — A short shaft or spindle on which an object may be mounted. Spindles or supports for milling machine cutters and saws are called arbors.

BASTARD — A coarse-cut file, but not as rough as a first cut.

BAUXITE — A white to red earthy aluminum hydroxide. It is largely used in the preparation of aluminum and alumina and for the lining of furnaces which are exposed to intense heat.

BLAST — The volume of air forced into furnaces where combustion is hastened artificially.

BORE — The internal diameter of a pipe, cylinder, or hole.

BORING — The operation of making or finishing circular holes in metal, usually done with a boring tool.

BUFFING — Finishing a surface using a soft cotton wheel or a belt with some very fine abrasive.

BULL GEAR — The large driving gear of the shaper.

BURR — A thin edge on a machined or ground surface left by the cutting tool.

BUSHING — A sleeve or liner for a bearing. Some bushings can be adjusted to compensate for wear.

CALIPER — A tool for measuring the diameter of cylindrical work or the thickness of flat work. There are inside and outside calipers.

CARBURIZING — The process of increasing the carbon content of low-carbon steel by heating the metal below the melting point while it is in contact with carbonaceous material.

CASEHARDENING — A process by which a thin, hard film is formed on the surface of low-carbon steel.

CEMENTED CARBIDE — A very hard metal carbide cemented together with a little cobalt as a binder, to form a cutting edge nearly as hard as a diamond.

CENTRE DRILL — A short drill, used for centring work in order that it may be supported by lathe centres. Centre drills are usually made in combination with a countersink, which permits a double operation with one tool.

CENTRE GAUGE — A gauge used to align a threading toolbit for thread cutting in a lathe.

CENTRE HEAD — A tool used for finding the centre of a circle or of an arc of a circle. It is most frequently used to find the centre of a cylindrical piece of metal.

CHAMFER — A bevelled edge or a cut-off corner.

CHATTER — Caused, while machining work, by lack of rigidity in the cutting tools or in machine bearings or parts.

CHIP PRESSURE — The force exerted on a cutting tool when removing material during machining.

COLD CHISEL — An all-steel chisel without a handle, used for chipping of metals.

COPPER — A soft, ductile, malleable metal, second only to silver for electrical conductivity. It is also the basis of all the alloys known as brass and bronze.

CORUNDUM — A natural abrasive material used in place of emery.

COUNTERBORE — A tool used to enlarge a hole through part of its length.

COUNTERSINK — A tool used to recess a hole conically for the head of a screw or rivet.

CRITICAL TEMPERATURE — The temperature at which certain changes take place in the chemical composition of steel during heating and cooling.

CROSSFEED — A transverse feed. In a lathe, that which usually operates at right angles to the axis of the work.

CUTTING SPEED — The speed in metres per minute or in feet per minute at which the tool passes the work, or vice versa.

DIE — An internal screw used for cutting outside thread on cylindrical work.

DOG — A clamp-type device that is fastened to work held between centres to connect the work to a positive drive in a milling machine or an engine lathe.

DRAW CHISEL — A pointed cold chisel, usually diamond shaped, used for shifting the centre of a hole being drilled to the correct location.

DRIFT — A strip of steel, rectangular in section, wedge-shaped in its length, used for driving drill sockets from their spindles or sleeves.

DRILL SOCKET — The socket which receives the tapered shank of a drill or reamer.

EMERY CLOTH — Powdered emery glued on cloth, used for removing file marks and for polishing metallic surfaces.

END MILL — A milling cutter, usually smaller than 25 mm (or 1 in.) in diameter, with straight or tapered shanks. The cutting portion is cylindrical in shape, made so that it can cut both on the sides and the end.

FACEPLATE — A circular plate for attachment to the spindle in the headstock of a lathe. Work can be clamped or bolted to its surface for machining.

FEED — The longitudinal movement of a tool in millimetres per minute or hundredths of a millimetre per revolution. Inch feeds are stated in inches per minute or thousandths of an inch per revolution.

FEELER — A gauge for determining the size of a piece of work, the accuracy of the test depending on the sense of touch. They are usually made in leaves and vary in thickness by hundredths of a millimetre for metric gauges and thousandths of an inch for inch gauges.

FERROUS METALS — Metals containing ferrite or iron.

FILLET — A concave or radius surface joining two adjacent faces of an article to strengthen the joint, as between two diameters on a shaft.

FINISHING — Machining a surface to size with a fine feed produced in a lathe, milling machine, or grinder.

FIXTURE — A special device designed and built for holding a particular piece of work for machining operations.

GAUGE — A tool used for checking dimensions of a job. A surface gauge can also be used to lay out for machining.

JIG — A device which holds and locates a piece of work and guides the tools which operate upon it.

KEYWAY — A groove, usually rectangularly cut in a shaft or hub, which is keyed and fitted for a driving purpose.

LEADSCREW — A threaded shaft which runs longitudinally in front of the lathe bed.

LIMITS OR TOLERANCE — The limits of accuracy, oversize or undersize, within which a part being made must be kept to be acceptable.

MANDREL — A shaft or spindle on which an object may be fixed for rotation, such as that used when a piece is to be machined in a lathe between centres.

MESH — The engagement of teeth or gears of a sprocket and chain.

METALLURGY — The art or science of separating metals from ores by smelting or alloying; the study of metals.

PARALLEL — A straight, rectangular bar of uniform thickness or width, used for setting up work in the same plane as a fixed surface.

PAWL — A pin having a pointed edge or hook, made to engage with ratchet teeth.

PERIPHERY — The line bounding a rounded surface, as the circumference of a wheel.

PILOT HOLE — A small hole drilled to guide and allow free passage for the thickness of the web of a twist drill.

PINION — The smaller of two gears in mesh, generally the driving gear.

PITCH — The distance from the centre of one thread or gear tooth to the corresponding point on the next thread or tooth. For threads, it is measured parallel to the axis. For gear teeth, it is measured on the pitch circle.

QUICK RETURN — The rapid movement of a part of a machine on the return stroke, such as a shaper ram or a milling machine table.

RACK — A straight strip of metal having teeth to engage with those of a gear wheel, as in a rack and pinion.

REAMER — A tool used to enlarge, smooth, and size a hole which has been drilled or bored.

ROUGHING — An operation done before finishing, to remove surplus stock rapidly where fine surface finish is not important.

S.A.E. (SOCIETY OF AUTOMOTIVE ENGINEERS) — These letters are used to indicate that the article or measurement is approved by the Society of Automotive Engineers.

SCALE — A thin surface on castings or rolled metal caused by burning, oxidizing, or cooling.

SERRATIONS — A series of grooves produced in metal to provide a grip or locking action. Vise jaws are often serrated.

SOAKING — Heating metal at a uniform heat for a period of time for complete penetration.

SPOTTING — An operation done in a lathe with a toolbit or a centre drill.

SURFACE GAUGE — A machinist's gauge consisting of a heavy base and a scriber for marking in layout for machining.

SURFACE PLATE — A cast iron scraped plate used for layout work. Granite plates are also used.

TENSILE STRENGTH — The resistance of steel or iron to a lengthwise pull.

TOLERANCE — The amount of interference required for two or more parts that are in contact. The amount of variation, over or under the required size, permitted on a piece of machined work.

TRUING — The operation of making the periphery of a job or tool concentric with the axis of rotation.

INDEX

Abrasive belt grinder, 238-40
Abrasive cloth, 149, 257
Abrasive cut-off saw, 86
Accuracy:
  of centre layout, 35-36
  of micrometer, 29
Adjustable wrench, 81
Alignment of lathe centres, 132-36
Alloy, 55
Alloy steels, 55, 57, 59
Alloyed castings, 52
Alloying elements in steel, 55
Alloys, nonferrous, 61
Aluminum, 61, 65, 79
Aluminum oxide abrasive belts, 238
Aluminum oxide grinding wheels, 232
Aluminum oxide straight wheel, 240
American National Acme Thread, 171
American National Thread, 171
Angle of thread, 173
Angle plates, 42, 43, 44, 207
Angular shaping, 196
Angular shoulders, 148
Annealing, 90, 258, 260
Apprenticeship training, 9, 64
Apron, lathe, 121, 122
Apron handwheel, 122
Arbor support, 201
Arbors, 204-05
Austenite, 257
Automatic crossfeed, lathe, 121, 122
Automatic feed mechanism, shaper, 189
Automation, 7

Babbage, Charles, 4, 248
Babbitt, 61
Ball-peen hammer, 65, 79
Bandsaws:
  contour-cutting, 89-94
  horizontal, 86, 87
  vertical, 89-94

Base of drill press, 97-98
Base of milling machine, 200
Basic oxygen process, steelmaking, 52, 53
Bastard cut file, 71
Bauxite, 61
Bearing surface, 109
Bed of lathe, 120
Bell centre punch, 41
Bench grinder, 7, 231-32
Bench-type drill press, 97
Bending steel, 260
Bessemer converter, 53
Billets, steel, 55, 56
Blade length, contour bandsaw, 92
Blade tension handle, 87
Blades:
  combination set, 38
  contour bandsaw, 90, 92
  cut-off saws, 86, 87-88
  hacksaw, 67
  screwdriver, 82
Blast furnace, 49, 50-52
Blind hole, 74
Blooms, steel, 55, 56
Blue vitriol, 36
Body, twist drill, 100-01
Boring, 115
Bottle car, 52
Bottoming tap, 74
Brass, 61, 65, 79
Brinell hardness test, 259
Brittleness, 57-58
Bronze, 61
Brown and Sharpe taper, 153
Bull wheel, crank shaper, 187
Bustle pipe, 51
Butting a rule, 22
Buttress saw band, 92
Butt welder, 90

C-clamps, 44
Calipers:
  firm-joint, 24
  hermaphrodite, 39, 41
  inside, 24-25
  micrometer, 26-30
  outside, 23-24, 145-46

  spring-joint, 24
  vernier, 30-32
Cam lock spindle nose, 131, 163, 165
Cape chisel, 69, 70
Carbon in steel, 56, 57, 257
Carbon steel drills, 100
Carburizing, 258
Care of tools, 20-21, 29-30, 71
Carriage, lathe, 121-22
Casehardening, 258
Cast iron, 52, 54, 55, 61
Castings, 52
Cemented carbide drills, 100
Centre drill, 108, 109-10
Centre head, 39, 40-41
Centre holes, drilling, 108-09
Centre lubricant, 136
Centre punch, 37, 41, 110
Centring work, 40-42
Chamfering, 74, 140, 176
Chilled iron castings, 52
Chip producing machines, 5
Chipping, 70, 208
Chisel edge, twist drill, 101, 110
Chisels, 69-70
Chromium in steel, 55, 57, 59
Chucks:
  cam-lock spindle nose, 131, 163, 165
  collet, 129, 130, 220, 221-22
  combination, 129, 130, 220
  drill, 99
  four-jaw independent, 129, 130
  taper spindle nose, 163, 164-65
  threaded spindle nose, 163, 164
  three-jaw universal, 129, 165-71
Clamp lever, 122
Clamps:
  C, 44
  parallel, 45
  shaper, 191
Clapper box, 187, 188
Claw saw band, 92

Clearance angle, 102, 127
Coal, 49, 50
Coarse cut file, 71
Code colour, 60
Coke, 50
Cold circular cut-off saw, 86, 87
Collet chuck, 219-20, 221-22
Collets, 129, 130, 220, 221-22
Column:
   drill press, 97, 98
   vertical mill, 219
Column face, 200
Combination chuck, 130
Combination set, 38-39
Combination square, 39
Compound rest, 121
Computers:
   history of, 247-49
   role of, 249-54
   use of, 2, 7, 13, 247
Computer-aided design (CAD), 2, 7, 249, 253
Computer-aided manufacturing (CAM), 7, 13, 249
Computer-integrated manufacturing (CIM), 7
Contact wheel, 240
Contour-cutting bandsaw, 90, 91
Contour surfaces, finishing, 94
Copper, 57, 61, 65, 79
Corrosion, 59
Counterboring, 80, 97, 114
Countersinking, 79, 80, 97, 108, 114
Crank shaper, 187
Crest of thread, 173
Cross filing, 73
Cross-slide, lathe, 121
Crossfeed direction lever, 188
Crossfeed handwheel:
   milling machine, 201
   surface grinder, 240
Crossfeed traverse shank, 188
Cupola furnace, 52
Cut-off saws, 86-87
Cut-off tools, 169
Cutting edges, twist drill, 101
Cutting fluids, 105-07
Cutting a groove, 152

Cutting off work in a chuck, 169-70
Cutting speeds:
   contour bandsaw, 93
   drill press, 103-05
   lathe, 124-25
   milling machine, 207-08
Cutting stroke, shaper, 194-97
Cutting tool overhang, 191
Cutting tools, 126-27, 190-91

Dead centre, 133, 136
Decimal inch system, 25-26
Depth of cut:
   drill press, 98, 104
   lathe, 126, 143
   milling machine, 209, 226
   shaper, 192
   surface grinder, 244
Depth of thread, 173
Depth stop, 173
Dial indicator, 134
Diamond dresser, 243
Diamond point chisel, 69, 70
Die stock, 78
Dies, 77-78
Direct indexing, 216-17
Divider, 37-38, 42
Dividing head, 214-17
Double-cut files, 61
Double-end wrench, 74, 81
Dowel pins, 80, 81
Downfeed handle, 188
Draw filing, 73
Dressing a wheel, 234-35, 243
Drift, 99
Drill chuck, 99
Drill gauge, 103
Drill grinding, 237-38
Drill point, 102
Drill press, 6, 96-116;
   feeds, 105
   operations, 114-16
   parts, 97-98
   speed, 103-05
Drill sizes, systems of, 102-03
Drill sleeves and sockets, 99
Drill vise, 112
Drilling:
   lathe centre hole, 108-10
   on a vertical mill, 227

pilot hole for large drills, 112-13
work fastened to an angle plate, 113
work fastened to a drill table, 112
work in a lathe chuck, 166-68
work in a V-block, 113
work in a vise, 111-12
Drilling head, 97, 98
Drills, 57, 97, 100
Drive plate, 162
Drop-forged parts, 59
Ductility, 58

Elasticity, 58
Electric furnace, 53, 55
Electro-chemical machines, 7
Electro-discharge machines, 7
Elevating screw, 201
End-relief clearance angle, 127
Endless abrasive belts, 86
Engine lathe, 6, 118;
   alignment of, 132-36
   and chucks, 162-71
   cutting speed and feeds for, 123-26
   cutting tools and holders, 127-28
   facing work on, 140-41
   filing in, 148-49
   function of, 119
   and grooving, 152
   and machining sequence, 138-49
   mounting work on, 136-38
   parallel turning on, 141-46
   parts of, 119-22
   and polishing, 149-50
   and safety, 122-23
   shoulder turning on, 146-48
   size of, 119
   and tapers, 153-62
   and thread, 171-81
   work holders, 129-32

Facing:
   work between centres, 140-41
   work in a chuck, 165-66

Fatigue failure, 58, 59
Feed change levers, 121
Feed dial, 201
Feed directional plunger, 122
Feed reverse lever, 121
Feed rod, 121
Feed:
   drill press, 105, 106
   engine lathe, 121, 125-26
   milling, 208-09
   shaper, 189
Ferrous metals, 56-57
File bands, 94
File card, 71
Files, 70-72
File Test, 259
Filing:
   flat surface, 244
   in a lathe, 148-49
   on a contour bandsaw, 93-94
Filleted shoulders, 147-48
Finish turning, 145
Finishing:
   contour surfaces, 240
   flat surfaces, 244
   holes, 227
Finishing cut, 190
Firm-joint calipers, 24
Fixture, 207
Flat cold chisel, 69, 70
Flexible rule, 22
Floor-type drill press, 97
Flutes, 100
Flycutter, 219, 220, 225-26
Forging steel, 260
Four-jaw independent chuck, 129, 130
Furnaces:
   basic oxygen, 54-55
   blast, 49, 50-52
   cupola, 52
   electric, 55
   heat treating, 255
   high-temperature, 255
   low-temperature, 255
   open hearth, 53-54
   pot-type, 255

Gas method, casehardening, 259
Gauge block, 29
Gauges:
   height, 42
   radius, 45
   surface, 42-43
   thread, 171
   toolbit grinding, 236
Gear cutter, 202
Gear shaper, 187
Geared-head lathe, 125
Graduated collars:
   milling machine, 209
   shaper, 192
Graduated micrometer collars, 142, 192
Gray iron castings, 52
Grinders, types of, 7, 231-32, 238-40
Grinding:
   drill, 237-38
   electrolytic, 7
   flat surface, 243-44
   finish, 239-40
   lathe toolbit, 235-37
   offhand, 232
Grinding wheels, 232-33, 242-43
Grooving, 152
Grooving chisels, 69

Hacksaw blade, 67
Hacksaws:
   hand, 67-69
   pistol grip, 67
   power, 86
Hammers:
   ball-peen, 65, 79
   soft-faced, 65
Hand feed lever, 98
Hand reaming, 78-79
Hand taps, 73-77
Hardening, 255, 259
Hardness of steel, 58, 59, 255, 259-60
Hardness tests, 259
Head, vertical mill, 219
Headstock of lathe, 121
Headstock spindle, 121
Heat colours, 258
Heat-treated steel, 255, 260
Heat treatment specifications, 260
Height gauge, 42

Hematite, 49
Hermaphrodite caliper, 39, 41
High-carbon steel, 57
High-speed steels, 57
High-speed steel drills, 100
High-strength steel, 57
High-temperature furnace, 255
Hold-downs, 191
Holes, tapping, 74, 76
Hook rule, 22
Hook-tooth saw, 92
Horizontal bandsaw, 87, 88
Horizontal spindle grinder, 240
Hot metal car, 52
Hydraulic shaper, 187

Inch measuring system, 19
Inch vernier calipers, 32
Index head, 214-17
Index head spindle, 214
Index plates, 214
Indexing, 214-17
Ingot moulds, 54, 55
Installing a saw blade, 87-88
Iron ore, 48, 49
ISO Metric Threads, 74-75, 172

Jacobs collet chuck, 130
Jarno taper, 153
Job selector, 90, 93, 94

Kasenit, 258
Keys, drill chuck, 99
Keyways, 213-14
Knee, 219
Knockout bar, 135
Knurling, 150-52

Ladle, 52, 55
Lathe apron, 121, 122
Lathe bed, 120
Lathe carriage, 121-22
Lathe centres:
   alignment, 132-36
   mounting and removing, 135
Lathe cross-slide, 121
Lathe dog, 136, 137, 176
Lathe file, 70
Lathes:
   engine, *see* Engine lathe

geared-head, 125
turret, 2
Layout, 35-36, 42-46
Layout dye, 36
Layout tools, 36-46
Lead, 61
Lead screw, lathe, 121
Lead of thread, 173
Letter size drills, 102
Limestone, 50, 51
Limonite, 49
Lip clearance, twist drill, 102
Liquid bath method, casehardening, 258-59
Live centre, lathe, 121, 129
Locating centres of round stock, 40-41
Low-carbon steel, 56
Low-temperature furnace, 255
Lower pulley, contour bandsaw, 90

Machine operator, 9-10
Machine screws, 79-80
Machine steel, 59, 77
Machine tools:
 evolution of, 1-3
 types of, 4-7
Machining:
 angular surface, 148, 197
 block square and parallel, 196, 222-25
 electro-chemical, 7
 electro-discharge, 7
 ends, 225-26
 flat surface, 195
 vertical surface, 197
Machinist, 2, 10-11
Machinist's hammer, 65
Machinist's vise, 65
Magnetite, 49
Malleability, 58
Malleable castings, 52
Mandrels, 131
Manganese in steel, 57
Manufacture:
 of cast iron, 52
 of pig iron, 50-52
 of steel, 52-56
Margin, twist drill, 101
Martensite, 257

Measurement systems, 19-20
Medium-carbon steel, 56
Metal fasteners, 79-81
Metal identification, 59, 60-61
Metal saw, 6
Metal shaper, 186-97
Metal stamps, 66-67
Metals:
 ferrous, 56-57
 nonferrous, 61
 physical properties of, 57
 shapes and sizes of, 61-62
Metric size drills, 102
Metric system, 19, 20
Metric vernier calipers, 30-31
Metricating lathe threading, 179-80
Micrometer calipers, 26-30
Micrometer parts, 26-27
Micrometers:
 inch, 28
 measuring with, 29-30
 metric, 30-31
 thread, 209
Mill file, 70
Milling:
 flat surface, 211-12
 hexagon, 215
 keyways, 213-14
 side, 213-14
 slots, 213-14
Milling cutters, 201-02, 205-06
Milling machines:
 cutting speeds of, 207-08
 feeds, 208-09
 function of, 199-200
 and graduated collars, 209
 operation of, 213-28
 parts of, 200-01
 safety with, 202-04
 setting up, 210-13
 types of, 6, 199-200
 vertical, 217-19, 220-28
Mining, 49
Molybdenum in steel, 57, 59
Morse taper, 153, 154
Mounting:
 arbor, 204
 chucks, 164-65
 file band, 94
 grinding wheel, 242-43
 milling cutter, 205-06

saw blade, 93
taper shank tools, 99
work between lathe centres, 140-41
work in shaper vise, 195

Necking, 152
New generation machines, 5-6
Nickel, 55, 57, 59, 61
Non-chip producing machines, 5
Nonferrous alloys, 61
Nonferrous metals, 61
Number size drills, 102
Number of threads, 172
Numerical control, 4, 7, 12-13, 250-54

Offhand grinding, 232
Offset screwdriver, 82
Open hearth furnace, 52, 53-54
Open pit mining, 49
Outside calipers, 23-24, 145-46
Overarm, 201, 219
Oxygen in steel making, 53, 54, 55

Pack method, casehardening, 258
Parallel clamps, 45
Parallel turning, 141-46
Parallels, 45, 191
Parting tools, 169
Parts of machines:
 contour bandsaw, 90-91
 drill press, 97-98
 horizontal bandsaw, 87, 88
 horizontal milling machine, 201
 lathe, 120-22
 shaper, 187-89
 surface grinder, 240
 vertical milling machine, 217-19
Pascal, Blaise, 248
Pedestal grinder, 7, 231-32
Pelletizing process, 49
Phillips screwdriver, 83
Phosphorous in steel, 57
Pig iron, 50-52
Pigs, 52
Pilot hole, 112-13

Pin spanner wrench, 82
Pinning a file, 71
Pistol grip hacksaw, 67
Pitch:
    micrometer thread, 27
    saw blade, 68, 87
    of thread, 173
Plug tap, 74
Point, twist drill, 101-02
Polishing, 149-50
Pot-type furnace, 255
Power hacksaw, 86
Precision saw band, 90
Preheating, 257
Prick punch, 37
Pyrometer, 256

Quenching, 257, 258
Quick-change gear box, 121, 125, 174
Quill:
    drill press, 98
    vertical milling machine, 219

Radius gauge, 45-46
Ram:
    shaper, 187-88
    vertical milling machine, 218
Ratchet stop, micrometer, 27
Reading a micrometer, 27, 28, 161-62
Reamer:
    hand, 78-79
    machine, 115
    steel in, 57
Reaming:
    hand, 78-79
    machine, 115
    on a vertical milling machine, 227
    work in a chuck, 168
Recessing, 152
Registration lines, 163, 165
Regular tooth saw band, 90
Removing:
    arbor, 204
    chucks, 162-63
    lathe centres, 135
    milling cutter, 206

Resetting a threading tool, 180
Ring gauge, 161, 181
Rivets, 79
Robertson screwdriver, 83
Rockwell hardness test, 259
Roller guide brackets, 87
Root of thread, 173
Rough turning, 138, 139, 143-45
Roughing cut, 120, 190
Roughing tool, 127
Round adjustable die, 77, 78
Round nose chisel, 69, 70
Round work, machining, 138
Rules, 21-23, 38

Saddle:
    lathe, 121
    milling machine, 201
Safety:
    abrasive belt grinder, 239
    bench or pedestal grinder, 232, 233
    chisels, 70
    cut-off saw, 93, 94
    drill press, 107-08
    engine lathe, 122-23, 152
    files, 71-72
    hammer, 66
    heat treatment, 256-57
    milling machine, 202-03
    shaper, 192-93
    surface grinder, 240-41
    thread cutting, 78
    wrenches, 82
Safety glass eyeshields, 16, 70, 232
Safety rules, 15-16
Saw bands, 90, 92
Saw blades, 87-88
Saw file, 94
Saw frame, 87
Saw guides, 90
Saw run-out, 88
Sawing:
    with hand hacksaw, 67-69
    hints on, 89
    to a layout, contour bandsaw, 93
    to length, 88-89

Saws:
    cut-off, 86-87
    metal, 6
    power, 85-94
Screwdrivers, 82-83
Screws, 79-80
Scriber, 36-37, 39
Scroll plate, 129
Second-cut file, 71
Sector arms, 214, 215
Self-holding tapers, 153
Self-releasing tapers, 153
Self-tapping screws, 80
Setting:
    cutter to work surface, milling machine, 211-12
    cutter to work surface, shaper, 195
    inside calipers, 24
    lathe feed, 126
    length of shaper stroke, 194
    outside caliper, 23
    position of shaper stroke, 194-95
    quick-change gear box for threading, 174, 175
    sector arms, 215
    surface gauge to a dimension, 42-43
    up the cutting tool, lathe, 124, 136
    up work for angular shaping, 197
    vise for angular shaping, 197
Shaft mining, 49
Shank, twist drill, 100
Shank end, tap, 99
Shapers:
    cutting tools for, 190-91
    feed mechanism of, 189
    function of, 6, 186-87
    mounting work on, 193
    parts of, 187-89
    safety and, 192-93
    setting stroke of, 194-95
    speeds of, 189-90
    types, 187
    work-holding devices for, 191
Shaping:
    angular surfaces, 196

flat surfaces, 195-96
vertical surfaces, 196
Sharpening:
  drill, 237-38
  toolbit, 235-37
Shell end mill, 202, 219, 220
Shore scleroscope test for hardness, 259
Short-length rule, 22
Shoulder turning, 146-48
Shoulders, 146
SI (metric system), 19
Side of thread, 173
Side milling, 213-14
Side-rake angle, 190
Side-relief clearance angle, 127, 190
Silicon carbide abrasive belts, 239
Silicon carbide grinding wheels, 232
Silicon in steel, 57
Simple indexing, 214
Single-cut files, 71
Single-end wrench, 81
Single-threaded worm, 214
Skip car, 50
Skip-tooth saw band, 92
Slabs, steel, 55, 56
Slag, 52, 55
Slag car, 52
Slots, 24, 213-14, 227-28
Smooth cut file, 71
Socket set screw wrench, 82
Soft contact wheel, 240
Solid die, 77-78
Solid end mill, 220, 225-26
Solid square, 38
Spark testing, 60-61
Special purpose machine tools, 7
Speeds:
  engine lathe, 123-25
  shaper, 189-90
  twist drills, 103-05
Spindle:
  drill press, 96
  index head, 214
  lathe, 136-37, 162-65
  milling machine, 154, 201, 220
  surface grinder, 240

Spindle binding lever, 122
Spindle sleeve, drill press, 98
Spindle speed dial, 201
Split-nut lever, 121
Spot-facing, 115
Spotting hole with centre drill, 110
Spring-joint calipers, 24
Spring-tempered rule, 21
Square head, 38-39
Square thread, 171
Squares:
  combination, 38, 43
  solid, 38
Squaring a shoulder, 146-47
Star dresser, 235
Steady rests, 132
Steel:
  alloy, 55, 59
  Chemical elements in, 57
  heat-treated, 255-60
  high-carbon, 57, 60
  high-speed, 57, 60
  low-carbon, 57, 60
  machine, 59, 77
  manufacture of, 52-56
  medium carbon, 57, 60
  tool, 257
  types of, 56-57
Steel rules, 38-39
Steel square, 38
Steep tapers, 153
Stencils, 66
Step pulleys, horizontal bandsaw, 87
Straightedge, 23
Straight-shank drill, 167
Straight-shank twist drills, 102
Strand casting, 56
Strip mining, 50
Stroke indicator, shaper, 189
Stroke regulator shaft, shaper, 189
Sulphur in steel, 57
Surface gauge, 42-43
Surface grinder, 240-42
Surface plates, 36
Surfaces, preparing for layout, 36
Swivel base, vise, 196, 197
Swivel table housing, 201

Table:
  contour bandsaw, 90
  drill press, 96, 98
  milling machine, 201, 219
  shaper, 188
Table feed mechanism, shaper, 189
Table handwheel, 201, 240
Table reverse dogs, surface grinder, 240
Table traverse handwheel, 240
Taconite, 49
Tailstock, 122, 132
Tailstock adjustment, aligning centres, 133-35
Tailstock handwheel, lathe, 122
Tailstock offset, 156-58, 159-61
Tailstock spindle, lathe, 122
Tang:
  file, 70, 72
  twist drill, 100
Tap drill, 75
Tap drill size, 75-77
Tap wrench, 75
Taper attachment, 158-59
Taper pins, 80, 81
Taper ring gauge, 159, 161
Taper shank drills, 99, 167
Taper spindle nose chucks, 151, 163, 164-65
Taper tap, 74
Taper turning:
  compound rest, 160
  offset method, 159-60
  taper attachment, 158-59
Tapered spindle nose, 131, 163, 164-65
Tapers:
  inch, 153, 155
  metric, 153-54, 155-56
  Morse, 153, 154
  self-holding, 153
  standard milling machine, 153
  steep, 153
Tapping:
  by hand, 76-77
  hole, 74, 75
  in drill press, 116
  in lathe, 170-71
Taps: